Biotechnology and Conservation of Cultural Heritage

Franco Palla • Giovanna Barresi
Editors

Biotechnology and Conservation of Cultural Heritage

Second Edition

Springer

Editors
Franco Palla
Department of Stebicef
University of Palermo
Palermo, Italy

Giovanna Barresi
Department of Stebicef
University of Palermo
Palermo, Italy

ISBN 978-3-030-97584-5 ISBN 978-3-030-97585-2 (eBook)
https://doi.org/10.1007/978-3-030-97585-2

This Springer imprint is published by the registered company Springer Nature Switzerland AG
The registered company address is: Gewerbestrasse 11, 6330 Cham, Switzerland

Foreword

Key to the work of conservation science is knowing more about the alteration processes induced by the metabolic activities of organisms on artworks, and thus being able to apply innovative biocleaning methods. Selecting the most effective restoration approach entails understanding the morphological and physiological characteristics of both micro and macro-organisms which may affect monuments, wall paintings, historical cellulosic materials, parchment, textiles, and even plastic objects.

The second edition of this book incorporates the results of new studies on the many living organisms that facilitate biological colonization and damage, and also deals in depth with cleaning procedures. Interestingly, the book reveals how nanotubes and their biocompatibility can support various conservation processes in the cultural heritage. By exploring alternatives to traditional materials and techniques, future perspectives are outlined. There is also a new chapter on the recovery and restoration of taxidermal finds of extinct Sicilian wolves. This represents a very good example of investigating the hidden treasures stored in museums throughout Italy.

This book provides scientists with the basic knowledge needed to understand the biological processes in the cultural heritage field and encourages conservators to adopt new cleaning procedures. This book will inspire undergraduate students to explore the ever-expanding fields of science to overcome new hurdles. It is also a unique tribute to the many scientists who have been involved in studies on biodeterioration and biocleaning. This captivating book reminds readers that science will continue to accomplish the unimaginable so long as people continue to submerge their minds in the artistic worlds of biology and chemistry.

Pisa University
Pisa, Italy

Maria Perla Colombini

Preface

Over the past few decades, biotechnology has provided innovative techniques useful for diagnosing cultural heritage deterioration induced by microbiological systems and for defining efficient conservation/restoration strategies. Seen from this perspective, the *International Congress on Molecular Biology and Cultural Heritage* held in Seville (Spain) in 2003 represents a milestone. This book comes from the experience gained in the last decade of basic and applied research, developed in the Laboratory of Biology and Biotechnology for Cultural Heritage (LaBBCH), within the framework of national and international research projects and in collaboration with other Italian and foreign research institutions.

In particular, biotechnology has found successful application in at least four areas in the conservative restoration of cultural asset, such as the characterization of *biodeterioration*, the analysis of *bioaerosols*, the sustainable application of *biocides*, and the development of innovative protocols for *biocleaning* and *bioremoval*.

A biotechnological approach to biodeterioration minimizes sample amount, contributing to understanding the contamination and complexity of microbial communities colonizing the cultural assets, as well as revealing unculturable species in both organic and inorganic substrates. This approach, based on genomic DNA analysis, has also proven useful in recognizing microbial systems in the aerosol of indoor cultural heritage environments, particularly for those representing a potential health risk for visitors and professionals.

It is well known that the removal of undesired layers can be performed by viable bacterial cells or purified enzymes (hydrolases), thus contributing to the development and definition of specific biocleaning/bioremoving protocols, opening up new perspectives for the biological removal of undesired layers.

Novel bioactive compounds from marine organisms have been employed to counteract microbial colonization, as well as plant extracts. Specifically, essential oil solutions, already utilized as a biocide in the food, medicine, and pharmaceutical industries are recently applied, as reported in this book, for controlling the microbial colonization of cultural heritage, representing a sustainable alternative to synthetic chemical biocides.

There is no doubt that biotechnology provides a plethora of information useful for setting up appropriate strategies that are totally safe for works of art, restorers, and environment, requiring a short time of application, straight to green conservation strategies in a sustainable restoration prospective. Case studies are reported in the Chap. 5 in order to demonstrate that a biotechnological approach may represent a valid alternative to traditional procedures generally used in the conservation/restoration of cultural assets.

Finally, a new chapter (Chap. 6) has been added, compared to the previous edition, in which biotechnological investigations are combined with conservative restoration procedures, in order to set up an integrated approach on peculiar specimens (taxidermal preparations) belonging to natural cultural heritage.

In this book, I have tried to present current knowledge in the field, highlighting the extraordinary power of DNA and of the novel bioactive molecules when applied to the conservative restoration of cultural assets.

I apologize to the authors of those papers who have escaped my attention, and also to those whose papers I have not quoted accurately.

I hope that the second edition, like the previous one, will prove to be a suitable text for those who work in the field of cultural and natural heritage conservation and adequate use, with the main purpose of transmission to future generations.

This book is dedicated to Roberta, Emanuele, and Federica, in exchange for the time I spent away from them.

Palermo, Italy Franco Palla

Acknowledgements

I am greatly indebted to the professionals in conservation of cultural heritage, colleagues, and students that I met during my 6 year as a Coordinator of the 5 years Degree in Conservation and Restoration of Cultural Heritage at the University of Palermo. Their constant and faithful collaboration thrusted me towards the search for new strategies for the sustainable conservation of cultural heritage. Particular thanks are directed to the *Assessorato Beni Culturali—Centro Regionale per la Progettazione e il Restauro*, to the *Soprintendenze BB.CC.*, to all the authorities and structures who institutionally perform the conservation and restoration of cultural assets in the Sicilian Region. I am especially grateful to the University of Palermo, CNR—IRIB Palermo Research Area, Enbiotech s.r.l., and all colleagues and professionals that resolutely participate in setting up biotechnological protocols.

The survey and the case studies were also carried out within the framework of the activity of the research project It@cha—Italian Technology for Advanced Applications in Cultural Heritage, "Ricerca e Competitività 2007–2013", PON 01_00625 (UNIPA).

Contents

Abbreviations

aDNA	Ancient DNA
AMPs	Antimicrobial Peptides
AST	Antimicrobial Susceptibility Testing
ATP	Adenosine triphosphate
BIOBRUSH	Bioremediation for Building Restoration of the Urban Stone Heritage
BLAST	Biophotonic Laser-Assisted Surgery Tool
BMA	Bioactive Molecule with Antimicrobial Activity
BME	Bioactive Molecule with Protease Activity
BMP	Bioactive Molecule with Esterase Activity
CEN	Comité Européen de Normalisation
CFU	Colony Forming Unit
CIMs	Culture-Independent Methods
CLSM	Confocal Laser Scanner Microscopy
CR	Control Region
DGGE	Denaturing Gradient Gel Electrophoresis
DNA	Deoxyribonucleic Acid
EAN	European Aeroallergen Network
EBML	European Molecular Biology Laboratory
EDTA	Ethylene-diamine-tetraacetic Acid
EOs	Essential Oils
ESEM	Environmental Scanning Electron Microscope
FEPA	Federal Environmental Protection Agencies
FISH	Fluorescence In Situ Hybridization
FTIR	Fourier Transform Infrared Spectroscopy
HKY	Hasegawa–Kishino–Yano
HNTs	Halloysite Nanotubes
HPD	Highest Probability Density
HPLC	High Pressure Liquid Chromatography
HVAC	Heating, Ventilation and Air Conditioning
IFSR	Regional Forest Inventory in Sicily

IMA	Index of Microbial Air
ITS	Internal Transcribed Spacer
LABBBC	Laboratory of Biology and Biotechnology for Cultural Heritage
LAMP	Loop-Mediated Isothermal Amplification
LD50	Lethal Dose 50%
LM	Light Microscope
MBC	Minimum Bactericidal Concentration
MCMC	Markov Chain Monte Carlo
MFC	Minimum Fungicidal Concentration
MIC	Minimal Inhibitory Concentration
ML	Maximum likelihood
MNPs	Marine Natural Products
MRCA	Most Recent Common Ancestor
mtDNA	Mitochondrial DNA
NAB	National Allergy Bureau
NCBI	National Center for Biotechnology Information
Nd:YAG	Neodymium-doped Yttrium Aluminium Garnet
NGS	Next-Generation Sequencing
NJ	Neighbour-joining
NRB	Nitrate-Reducing Bacteria
OM	Optical Microscopy
PBAP	Primary Biological Aerosol Particle
PBAPs	Primary Biological Aerosol Particles
PCR	Polymerase Chain Reaction
RAPD	Random Amplification of Polymorphic DNA
REA	Red Española de Aerobiología
RH	Relative Humidity
RIMA	Rete Italiana di Monitoraggio in Aerobiologia
RNA	Ribonucleic Acid
RNSA	Réseau National de Surveillance Aérobiologique
rRNA	Ribosomal Ribonucleic Acid
SAS	Surface Air System
SBAP	Secondary Biological Aerosol Particle
SBAPs	Secondary Biological Aerosol Particles
SDA	Sabouraud Dextrose Agar
SEM	Scanning Electron Microscope
SEM-BSE	Scanner Electron Microscope Backscattered Electrons
SEM-EDS	Scanning Electron Microscopy with X-Ray Microanalysis
SNPs	Single Nucleotide Polymorphisms
SRB	Sulphate-reducing bacteria
SRY	Sex Determining Region Y
TE	Tris-EDTA buffer
TEM	Transmission Electron Microscope

TRIS-HCL	Tris(hydroxymethyl)aminomethane
TSA	Tryptic Soy Agar
UV	Ultraviolet
VPPS	Volumetric Pollen and Particle Sampler
XRF	X-Ray Fluorescence Spectroscopy

Biodeterioration

1

Enza Di Carlo, Giovanna Barresi, and Franco Palla

Abstract

The biodeterioration of organic and inorganic materials, as well as polymers, is a complex of alteration processes induced by the growing and metabolic activity of organisms. It can be recognized on monuments, wall paintings, stone, wood, paper, vegetal/animal fibers, and parchment artworks. As defined by Hueck, 1968, biodeterioration is "any undesirable change in the properties of a material caused by the vital activities of organisms"; this definition is accepted as the meaning of the phenomenon. Both macroorganisms (such as animals, plants, and mosses) and microorganisms (such as autotrophic or heterotrophic bacteria, microfungi, cyanobacteria, algae, and lichens) represent the triggers of biodeterioration for cultural heritage. Understanding the morphological and physiological features of the biodeteriogens is required to establish the kind of interaction that occurs with the material and to assess the cause-effect of the biodeterioration action of a specific identified biological agent. For a complete evaluation of biodeterioration, a proper sampling and identification of the majority of biodeteriogens are required. Therefore, in order to apply a prompt and effective conservation to limit further damage, evaluating and quantifying the presence of biological systems that induce damage in heritage materials is indispensable.

Keywords

Cultural heritage material · Wall painting · Microbial colonization · Stone surface · Sampling and culture methods

E. Di Carlo (✉) · G. Barresi · F. Palla
Department of STEBICEF, Laboratory of Biology and Biotechnologies for Cultural Heritage, University of Palermo, Palermo, Italy
e-mail: franco.palla@unipa.it

F. Palla, G. Barresi (eds.), *Biotechnology and Conservation of Cultural Heritage*,
https://doi.org/10.1007/978-3-030-97585-2_1

1.1 Biological Systems as Deteriogens

An object can represent a *microcosmos* in which several types of vital organisms—and their metabolic excretes or products—coexist, in response to climatic or environmental parameters and to chemical or physical materials properties. Both artifacts of artistic-historical value stored in museums or galleries, and monuments or historical buildings exposed outdoors, are largely subjected to deterioration processes by macro- and microorganisms (e.g., rodents, birds, plants, insects, lichens, algae, bryophytes, fungi, bacteria, cyanobacteria) commonly known as *biodeteriogens*. Biological systems are able to initiate, support, and accelerate some chemical and physical reactions, representing a serious threat for the integrity and conservation of cultural assets, in both indoor and outdoor sites (Dakal and Cameotra 2011).

The biodeterioration of a wide range of cultural heritage materials has been largely investigated, and the effects of biodeterioration caused by living organisms are well documented (Hueck 1968; Lombardozzi et al. 2012; Piervittori et al. 1996; Pyzik et al. 2021). In relation to nutritional requirements and metabolic properties, organisms show different responses to colonized materials and can cause different types of damage; knowledge of the biodeteriogens, their nutritional requirements, their growing settings, and their manifestation once they have colonized cultural items is indispensable to recognize the severity of the biodeterioration process.

The biodeteriogens' varieties should be grouped according to their attendance in different materials and to their type of damage: Table 1.1 shows the main groups of biodeteriogens.

In biological colonization, there are always some pioneer cells that facilitate the colonization of other organisms, especially in those materials (e.g., stone) that do not support the growth of heterotrophic organisms but can support food and organic energy sources for them in the form of other microbes, insect fragments, and bird droppings (Sharma et al. 1985; Hochella 2002; Vaughan et al. 2002; Warscheid and Braams 2000).

The bioaerosol is also a potential source of biodeteriogens, due to the large number of biological particles suspended in the air (fungal and bacterial spores, lichen and algal cells, pollen grains); the spores transported by air or dust can represent a threat of paper and graphic collections (books, photographs, etc.) preserved in libraries, archives, and repositories. Throughout, these locations are generally characterized by poor ventilation, and the absence of periodic cleaning can compromise the integrity of stored objects and in addition can facilitate the proliferation of dangerous fungi (Caneva et al. 2020). An object evidently has different levels of susceptibility to biodeterioration or colonization, also depending on human contribution in terms of care of cultural heritage.

Which are the main actors that induce biodeterioration? The following biological systems have been largely known to induce biodeterioration phenomena on cultural heritage, in both organic and inorganic materials.

Table 1.1 Major biodeteriogens affecting cultural heritage materials and main related physical and chemical damages

	Physical damage	Chemical damage	Material
Higher plants	Cracks, detachment of stone	Roots excrete organic acids	Natural and artificial stones
Animals and insects	Holes, losses, and surface erosion and disintegration, structural damages	Droppings and urine	Stone, wood, paper, parchment and leather, vegetal and animal fibers, and more
Mosses and liverworts	Physical intrusion by rhizoids	Extraction of mineral from substratum, production of carbonic acid	Natural and artificial stones
Lichens	Cracks and fissures	Releasing of highly corrosive organic acids	Limestone, sandstone
Algae and cyanobacteria	Powdering	Staining, disintegration of stone	Natural and artificial stones, wall paintings, wood
Fungi	Contraction and expansion of thallus, fissures, loss of materials	Production of organic and inorganic acids and pigments, chelating properties	Stone, wall paintings, wood, paper, parchment and leather, vegetal and animal fibers, and more
Heterotrophic bacteria	–	Staining, production of acids and pigments	Stone, wall paintings, wood, paper, parchment and leather, vegetal and animal fibers, and more
Autotrophic bacteria	–	Production of organic and inorganic acids, biofilm formation, staining	Stone, wall paintings, wood, paper, parchment and leather, vegetal and animal fibers, and more

1.1.1 Macro-systems

Higher plants pose a severe threat especially for the conservation of archaeological remains and buildings, as they can involve physical and chemical diseases (Celesti-Grapow and Blasi 2004; Mishra et al. 1995). The damage caused by plants on monuments or archaeological sites is well known and is linked either to particular biological forms or to the characteristics of root systems (Caneva et al. 1994). Archaeological sites are characterized by rich flora and vegetation (Caneva et al. 2003; Capotorti et al. 2013) where the plants can develop in various structures (e.g., vertical/inclined wall surfaces, horizontal surfaces) and in different ecological conditions (Fig. 1.1). Lisci et al. (2003) described two types of colonization; the first regards an attack on the wall structure, usually the result of abiotic factors that create suitable conditions for microbial growth (bacteria, fungi, and lichens), which accelerate the deterioration of the structure and lead to the formation of a substrate for the germination of seeds. It is well known that some wall plants, especially perennial ones, with their radical apparatus and biomass can compromise the

Fig. 1.1 Higher plant massive growth on Cambodian temple (Angkor Park)

integrity and structural stability of ancient monuments (Mishra et al. 1995; Caneva et al. 2006). The second mode of colonization ensues mostly on horizontal surfaces with a water supply; the pioneer plants are mosses that trap atmospheric dusts, allowing the formation of appropriate substrate for the germination of other plants. Moreover, higher plants with a root system may penetrate deep into the structure and grow, causing physical and chemical damage, through the exudation of organic acids. The deep penetration of the roots of tree species (such as *Pinus pinea* and *Quercus ilex*) can be dangerous for hypogenous structures, producing the detachment of plaster and the collapse of walls, their mechanical force can open cracks, cause crumbling, loosen stones and large fragments of the wall (Tiano 1986; Bettini and Villa 1976; Caneva and Galotta 1994; Lisci et al. 2003; Almeida et al. 1994).

Mosses and liverworts (bryophytes) are able to penetrate some types of stone with their rhizoids. The damage they cause is mainly associated with aesthetic appearance (usually green-gray patches). The capacity of the mosses to accumulate calcium ions from the substratum is associated with their biodeterioration capacity; the carbonic acid produced as a result of their cellular respiration can cause damage to stone. Moreover, the death of mosses causes indirect damage to monuments and stones by enriching and increasing the humus content which supports the growth of successive higher plant species (Dakal and Cameotra 2012).

As regards the role of animals in the biodeterioration of cultural heritage, direct and indirect actions can be distinguished; for example, the direct action of birds is both physical, caused by crushing and scraping, and chemical, caused by the release of acid excrements containing high amounts of nitrate and phosphate compounds. Indirect damage occurs as a result of organic substances accumulated on stone surfaces, which can serve as nutritive substrata for heterotrophic microflora.

Molds affecting archival and library materials can display different types of interaction with insects that feed on the materials of stored objects, and the biological particles can be vehicle by visitors or, sometimes, by synanthropic rodents, eating wood, leather, and other soft organic materials. Many insects are able to damage wood using it as a nutrient source and for shelter and egg laying. The way adult insects or larvae digest cellulosic compounds varies (Allsopp et al. 2004). Among insects, silverfishes are a threat to any heritage collection; in particular, they eat glue and starch in books and in paper objects; termites eat cellulose-based materials (e.g., wooden furniture and artifacts, textiles, ethnographic artifacts); beetles can cause extensive damage to a variety of proteinaceous materials (e.g., wool, feathers, silk) and cellulosic textiles.

Furthermore, submerged or waterlogged materials, such as ceramics and wood, can suffer damage by several marine organisms that use them as a support for their growth. Depending on their chemical composition, the submerged stone artifacts are susceptible to the action of corrosion by perforating animal and plant organisms, such as sponges and bivalves that can induce macroboring (Ricci and Davidde 2012; Casoli et al. 2015).

1.1.2 Micro-systems

The term "microorganism" covers a wide variety of life forms, including bacteria, cyanobacteria, algae, lichens, and fungi. All cited microorganisms have different ecological properties and cause different damage in organic and inorganic materials of cultural heritage.

Favorable environmental conditions and the presence of energy sources with organic or inorganic nutrients allow the biological settlement of an exposed surface (Urzì and Krumbein 1994). The consequence of the biological activity is the formation of biofilms, colored patinas, encrustations and the presence of vegetative and reproductive bodies. In addition, fungi and bacteria are commonly isolated from libraries, archives, and museums' collections. Moreover, in indoor environments, the high presence of certain microbial species may involve irritation of the eyes and respiratory tract and can induce headaches, drowsiness, skin rash, and itching of the skin (Green and Scarpino 2002).

1.2 Microbial Colonizers

Microorganism biodiversity includes bacteria, archaea, and eukaryotes, which are extraordinarily diverse in their requirements for growth, and their proliferation is greatly affected by the nutrients available in their environment. Different kinds of microorganisms colonize artworks: chemoheterotrophic bacteria, chemolithotrophic bacteria, phototrophic bacteria, algae, and fungi. In the case of microbial-induced biodeterioration, alteration processes can arise when environmental conditions are favorable to growth (Nielsen et al. 2004; Nittérus 2000; Caneva et al. 1994) or when the existence of nutrients favors the colonization (Saiz-Jiménez 1993; Sterflinger and Piñar 2013). Microbial development depends on a combination of factors including relative humidity (RH, %), temperature fluctuation, natural or artificial lighting, moisture content, dust content, osmotic pressure, and carbon dioxide concentration in the atmosphere (Valentin 2003; Woese 2000; Caneva et al. 2008).

1.2.1 Algae and Cyanobacteria

Among *micro*-biodeteriogens, photosynthetic ones are potentially the most aggressive, due to their ability to develop on stone surfaces, causing colored patinas and incrustations (Tomaselli et al. 2000). Algae and cyanobacteria are usually the first colonizers of historical monuments due to their photosynthetic nature, causing aesthetic damage and indirectly supporting the growth of other microorganisms (Urzì and Krumbein 1994). These organisms can develop on exposed stone when a suitable combination of dampness, warmth, and light occurs and inorganic nutrients (e.g., calcium and magnesium minerals) are present. Algae may also cause biochemical deterioration, producing small quantities of organic acids (they dissolve and powder the stone), proteins, and sugars, which can promote the growth of bacteria.

Cyanobacteria are organisms traditionally included among algae, but they have a cell structure typical of bacteria. Cyanobacteria are a morphologically diverse group of phototrophic prokaryotes with the ability to synthesize *chlorophyll a* and *phycobilin* pigments (Sobiechowska-Sasim et al. 2014). Depending on the kind of organism and on the cycle phase, dark-green-, brown-, gray-, and pink-colored patinas may occur (Huer 2008). All cyanobacteria are unicellular, though many grow in colonies or filaments, often surrounded by a gelatinous or mucilaginous sheath.

The photosynthetic activity of these microorganisms enriches the substrate with organic carbon in various forms, and subsequently, the growing biomass entraps dust and soil particles providing further nutrient enrichment (Hirsch et al. 1995; Souza-Egipsy et al. 2004). The tolerance to desiccation, ability to utilize efficiently low light intensity and tolerate high levels, and resistance to high temperatures are important features of cyanobacteria that explain their widespread occurrence (Lamenti et al. 2003). Cyanobacteria have the ability to survive under drying and rehydration conditions on exposed monument surfaces and can protect themselves from harmful UV radiation by pigment production; they are also recognized in hypogeal environments (Albertano et al. 2003; Hoffmann 2002; Roldan et al. 2004).

1.2.2 Bacteria

Bacteria are a group of prokaryotic unicellular or colonial organisms of various shapes, classified into five groups according to their shapes: cocci, bacilli, spirilla, vibrios, or spirochaetes. The bacterium has a fairly thick cell wall made of peptidoglycans (carbohydrate polymers cross-linked by proteins); such bacteria retain a purple color when stained with a dye known as crystal violet and are known as Gram-positive (staining procedure). Other bacteria have double cell walls, with a thin inner wall of peptidoglycan and an outer wall of carbohydrates, proteins, and lipids. Such bacteria do not stain purple with crystal violet and are known as Gram-negative (La Placa 2005).

Bacteria involved in the deterioration of monuments and artworks mainly belong to three nutritional groups: photoautotrophs, chemolithoautotrophs, and chemoorganotrophs. Various species in photoautotrophs and chemolithoautotrophs groups (cyanobacteria, sulfur-oxidizing, and nitrifying bacteria) produce strong inorganic acids (sulfuric and nitric, respectively). Chemolithoautotrophic bacteria can derive their energy from the oxidation of reduced inorganic substances, using carbon dioxide as main carbon source. Endolithic nitrifying bacteria are the main representatives of the chemolithoautotrophic microflora in building stones; Kauffmann (1952) and Wagner and Schwartz (1965) first indicated the significance of nitrifying bacteria for biodeterioration.

Chemoorganotrophic bacteria produce several organic acids that can solubilize the mineral components of stones, such as the genera *Flavobacterium*, *Pseudomonas*, and *Microbacter* (Tiano and Tommaselli 1989; Tayler and May 1995; Swings and Descheemaeker 1995; Ortega-Calvo et al. 1993). Among the microorganisms dwelling on stone monuments, the autotrophic ones are considered the pioneering inhabitants.

Biodeterioration of organic substrates is a process involving several types of bacteria and represents a severe problem for archives, museums, and libraries. Bacteria also display a wide diversity in enzyme production, including lipases, proteases, and oxidoreductases (Neelakanta and Sultana 2013). Intense research on advanced microbiological systems based on the use of microorganisms for the removal of alterations on works of art has shown them to be a viable alternative for cultural heritage restoration (Ranalli and Sorlini 2008; Palla et al. 2013, Martino et al. 2015; Barresi et al. 2015; Parulekar-Berde et al. 2020).

1.2.3 Fungi

Fungi are usually classified in four divisions: the *Chytridiomycota*, the *Zygomycota*, the *Ascomycota*, and the *Basidiomycota*. There are also two conventional groups that are not recognized as formal taxonomic groups; these are the *Deuteromycota* and lichens. *Deuteromycota* include all fungi which do not reproduce themselves sexually, and lichens are a group of composite organisms formed by the association of algae and fungi. Frequently associated with biodeterioration of rocks and outdoor stone, lichens are highly resistant to desiccation and extreme temperatures. They are able to produce pigments, such as chlorophyll, carotenoid, and melanin, that may generate chromatic variations toward yellow, orange, red, or even brown (Krumbein 2003; Tiano and Tommaselli 1989; May et al. 1993; Palla et al. 2010).

Fungi are metabolically more versatile than other biodeteriogens because they are able to colonize on a wide variety of substrates (Sterflinger 2010; Onofri et al. 2014). They can exert both a biomechanical action through the disaggregation and reaggregation of the mineral fraction of stones and a biochemical action by producing metabolic organic and inorganic acid and by the absorption of metals by mycelia felts (Burford et al. 2003; Sterflinger 2000). Due to their enzymatic activity, fungi are also able to inhabit and to cause decay in paintings, textiles, paper, parchment, leather, oil, casein, glue, and other materials used for historical-artistic objects. Several fungal metabolic products are strongly colored, and the phenomenon of staining produced on the damaged areas on monuments can be the consequence of fungal development. In indoor environments (e.g., archives), when relative humidity increases and no adequate ventilation occurs, the conidia may be deposited more quickly on documents and deteriorate the document supports (Borrego et al. 2012).

Fungal flora inside environments are representative of the outdoor atmosphere, since airborne spores penetrate through doors and windows and the spores in the airborne particulate can cause an increase in allergies and respiratory and skin diseases, headaches, asthma, and weariness (Borrego et al. 2012).

1.3 Microbial Metabolic Activities and Deterioration of Cultural Assets

Both in organic and inorganic matters, evidence of biodeterioration is related to physical and chemical processes induced by microorganisms; physical damage is generally induced by growth or penetration within the material with the formation of micro-fractures, loss of cohesiveness, and disaggregation of the substrate (López-Miras et al. 2013; Cataldo et al. 2005). Physical damage caused by microbial colonization is less extensive than chemical damage, which can arise when products of metabolic activity interact with the material, causing alteration of the substrate, which at times is irreversible. Biochemical deterioration induced by microbial colonization consists both in digestion processes, when microorganisms use the substrate as nourishment, especially when organic compounds are utilized, and in metabolic excretions, waste products, or other substances (e.g., organic and inorganic acids, pigments) (Strzelczyk 2004; Saiz-Jimenez 1999; Perry et al. 2003). Many factors can contribute to biodeterioration, including the properties of the original materials, sources of nutrients, and the environmental conditions in which the objects are stored or exposed (Warscheid and Braams 2000; Prieto et al. 2006;

Pavlogeorgatos 2003). However, the complete understanding of biodeterioration processes is difficult, being the result of complex microbial interactions (Mandrioli et al. 2003; Warscheid and Braams 2000).

1.3.1 Inorganic Substrates

Artworks made of inorganic materials are exposed to natural processes of biodeterioration, especially when they are located outside (Cutler et al. 2013; Kumar and Kumar 1999; Dakal and Cameotra 2012). The biodeterioration process affects monumental stones, statues, historical buildings, wall paintings, archaeological remains, and, to a lesser extent, glasses and metals (Giustetto et al. 2015; Biswas et al. 2013; Gorbushina and Palinska 1999; Gorbushina et al. 2004; Piñar and Sterflinger 2009).

Numerous parameters influence the succession of microorganisms on stone; firstly, the properties of the stone itself determine the colonization pattern. The mineral composition, structure-texture, porosity, and permeability of stone may influence the distribution of such organisms in the monuments (Miller et al. 2006). Favorable environmental conditions and the presence of nourishment sources allow the biological colonization of an exposed stone surface (Miller et al. 2000; Palla et al. 2003; Anagnostidis et al. 1991). It is known that biological growth on stone is highly dependent on climatic and microclimatic conditions, such as humidity, temperature, light, and atmospheric pollutants (Moroni and Pitzurra 2008; Mansch and Bock 1998; Zanardini et al. 2000; Nuhoglu et al. 2006). To avoid microbial proliferation, it is necessary to control environmental factors, which becomes difficult in archaeological sites or in urban spaces but is more easily achieved in closed environments where the control of climatic and microclimatic conditions is easier (Salvadori and Charola 2011). Occasionally, the metabolic activities of microorganisms (autotrophic and heterotrophic) induce different types of damage: physical, when pressure is exerted by the growth of vegetative structures (e.g., lichenic and fungal thalli); chemical, when the excretion of enzymes, the production of inorganic and organic acids, and the liberation of chelating compounds occur; and aesthetical, as in the effect of releasing pigments (colored patches or patinas) (Ascaso and Wierzchos 1995; Ortega-Calvo et al. 1995; Albertano and Urzì 1993). Microbial colonization of cultural stone heritage may cause aesthetic changes due to the growth of pigmented microorganisms; cyanobacteria, algae, fungi, lichens, and some pigmented bacteria are responsible for these effects (McNamara and Mitchell 2005; Crispim and Gaylarde 2005). The algae and cyanobacteria can be considered the pioneering colonizers of a stone surface (Tomaselli et al. 2000), and as a result of their development, green-, brown-, and gray-colored patinas may also occur (Fig. 1.2) (Hauer et al. 2015; Golubíc et al. 2015). Lichens can be identified by directly observing the stone surface and are regarded as major biodeterioration agents in outdoor stone monuments (de los Ríos et al. 2009; Chen et al. 2000). Lichens and fungi can cause serious degradation by physical penetration: fungal hyphae are able to penetrate deeply beneath the stone surface, contributing to mechanical

Fig. 1.2 Differently pigmented biological systems (algae and cyanobacteria) colonizing wall surface in hypogeal environment

deterioration. This penetration simultaneously allows the transport of water and nutrients through the stone, facilitating internal colonization (Fernandes 2006).

Many microorganisms play an important role in the deterioration of stone often through the action of organic and inorganic acids produced as metabolic products (Adamo and Violante 2000; Sazanova et al. 2014). Generally, most microorganisms excrete organic acids while metabolizing organic and inorganic compounds. Organic acids react with the substrate through the action of protons and chelation of metal ions. The excretion by the lichen of low-molecular-weight organic carboxylic acids, such as oxalic, citric, gluconic, and lactic acids, with combined chelating and acidic properties, is a phenomenon of high intensity (Adamo and Violante 2000).

The production of inorganic acids is limited to some systematic groups, in particular to sulfobacteria and nitrobacteria; due to the action of nitrous and nitric acid excreted by nitrifying bacteria (*Nitrosomonas* spp. and *Nitrobacter* spp.), as well as sulfuric acid by sulfur-oxidizing bacteria (*Thiobacillus* spp.), stone dissolution and the formation of nitrate salts of stone (biocorrosion) can ensue (Fernandes 2006; Mansch and Bock 1998; Bock et al. 1990). Endolithic nitrifying bacteria are the main chemolithoautotrophic microflora in building stones. On the other hand, heterotrophic microorganisms develop on inorganic materials if nutrients, e.g., settled particulate matter, organic residual substances, and organic resins, glues, or binders by restoration practices, have been deposited on them (Caneva et al. 1998).

The fungal genera which have been demonstrated to be more abundant on inorganic materials, especially on stone, are *Cladosporium*, *Penicillium*, *Trichoderma*, *Fusarium*, and *Aspergillus* (de la Torre et al. 1993; Sazanova et al. 2014; Torre et al. 1993; Gu et al. 1998; Hu et al. 2013; Prasad Gupta and Sharma 2012; Sharma et al. 2011). The reduction and oxidation of mineral cations are characteristic activities of fungi (de la Torre and Gomez-Alarcon 1994). Several fungal metabolic products are strongly colored; dark spots are attributed to the presence of fungi of the family of *Dematiaceae*, which contains melanin pigments inside the mycelium. Black microcolonial fungi (MCF) play an important role in structural and aesthetic alteration and are considered among the most harmful microorganisms associated with monumental stone biodeterioration (Marvasi et al. 2012; Wollenzien et al. 1995; Isola et al. 2016; Salvadori and Municchia 2016). The dark gray and black crusts and patinas generally found on stone monuments, in particular, calcareous stones, can be derived by secreted dark compounds such as melanins and/or humic acids (Delgado Rodrigues and Valero 2003).

The organisms are often organized as microbial biofilms that cover the surface of the material and/or penetrate the substratum (Hoppert and König 2006). Biofilms often excrete polysaccharides, lipids, pigments, and proteins (Hauer et al. 2015). Through in situ microscopy, the zones where biodeterioration processes take place can be detected as biofilms composed of different microorganisms; endolithic microbial colonization seems to be induced by enhanced moisture availability in cracks and fissures (de los Rìos and Ascaso 2005). The enrichment of sulfur- and hydrocarbon-utilizing bacteria in the biofilms may contribute to dissolution of the stone (Mitchell and Gu 2000).

Detection and identification of damage induced by microorganisms' activity within inorganic substrates is not easy. Anyway, some researchers are using several methodologies and integrated techniques in order to recognize the different microorganisms and their associated biodeterioration (de los Ríos et al. 2009; Suihko et al. 2007; Gutarowska 2010; Fernandes 2006; Herrera et al. 2009).

1.3.2 Organic Matter

Organic-based materials of cultural heritage (e.g., paper, fibers, wood, papyri, leather and parchment) are generally subject to the aggression of several heterotrophic microorganisms that use them as a source of nourishment (Strzelczyk et al. 1997; Manente et al. 2012; Kavkler et al. 2015; Sterflinger 2010; Vukojević and Grbić 2010). Additionally, the presence of organic residues (e.g., glue, dirt, dust) may accelerate the processes of degradation, such as the loss of strength and elongation, oxidation, discoloration, and breakdown of molecular structures (Guiamet et al. 2014).

Most microorganisms are specialized in producing hydrolytic or proteolytic enzymes, necessary for the degradation of cellulose or collagen, respectively (Strzelczyk 2004; Sterflinger and Pinzari 2012). The production of extracellular enzymes and the extraction of aggressive metabolic products increase the loss of material (López-Miras et al. 2013). Enzymatic activity (due to cellulases, glucanases, laccases, phenolases, keratinases) plays an important role in the decay of organic materials, especially for libraries and archives heritage; moreover, if collections are exposed to high humidity, high temperature, and insufficient air circulation, fungal colonization is more dangerous (Sterflinger 2010; Rakotonirainy et al. 2015; Guiamet et al. 2014). Fungi often cause serious aesthetical spoiling and chromatic alteration due to the formation of colonies and fungal pigments. When they grow on paper, they degrade all its carbon-containing components such as cellulose by excreting enzymes or organic acids (oxalic, fumaric, succinic, and acetic acids), which settle over the substrate and acidify (López-Miras et al. 2013; Canhoto et al. 2004; Kavkler et al. 2015). Microscopy is a very useful technique in revealing the biological structures related to the development of hyphae in cellulolytic fungi (Guiamet et al. 2014). In paper, leather, or fibers, microbial growth mainly appears as patches of different colors (purple, yellow, brown, black, red, and green), shapes and sizes, related to the presence of pigmented fungal mycelium and spores or to exo-pigments produced by bacteria and fungi (Fig. 1.3a, c).

Paper documents have received particular attention in the past, mainly for a staining phenomenon called *foxing*. Foxing, a typical phenomenon in paper or in textile, is the name of random circular and irregular yellowish to brownish-red stains, fluorescent under ultraviolet rays, on the surface of old books, documents, maps, etc. (Montemartini Corte et al. 2003; Kraková et al. 2012; Rakotonirainy et al. 2015). The role of fungi, and whether they accompany the formation of stains, is still not clearly understood. Many authors dealing with the phenomenon of foxing on paper point out its microbiologic origin, but, in fact, foxing-causing fungi can rarely be cultured. Karbowska-Berent et al. (2014) showed foxing as well as numerous hyphae and fluffy coatings and were able to isolate five strains, belonging to the *Eurotium*, *Aspergillus*, and *Penicillium* genera. Due to their enzymatic activity (exo-enzymes such as cellulases, glucanases, laccases, phenolases, keratinases, monooxygenases), fungi are able to inhabit and decay not only paper heritage but also paintings, textiles, parchment, oil, casein, glue, and other materials used for historical art objects.

In paper decay, the most numerous bacteria are heterotrophs (*Cytophaga*, *Cellvibrio*, and *Actinomyces*). Deterioration by bacteria and fungi leads to loss of strength of the natural fibers, causing odor emissions, aesthetic damage, staining, and loss of fiber structure. Fungal damage of textiles can result in discoloration, staining, and smell due to the production of volatile compounds and the enzymatic and mechanical degradation of the material through the activity growth of fungi (Guiamet et al. 2014; Kavkler et al. 2015). Other important properties of fungi are related to their pathogenicity for workers involved in collection maintenance (Borrego et al. 2010; Sahab Ahmed et al. 2014).

More complex biodeterioration process occurs in composite materials, such as paintings on canvas, where often it starts on the reverse side, due to the presence of support polymers and glue; these components can be substrates for microbial growth. On the other hand, the organic materials present on the *recto* of the paintings are susceptible to attack by specialized microorganisms and by occasional

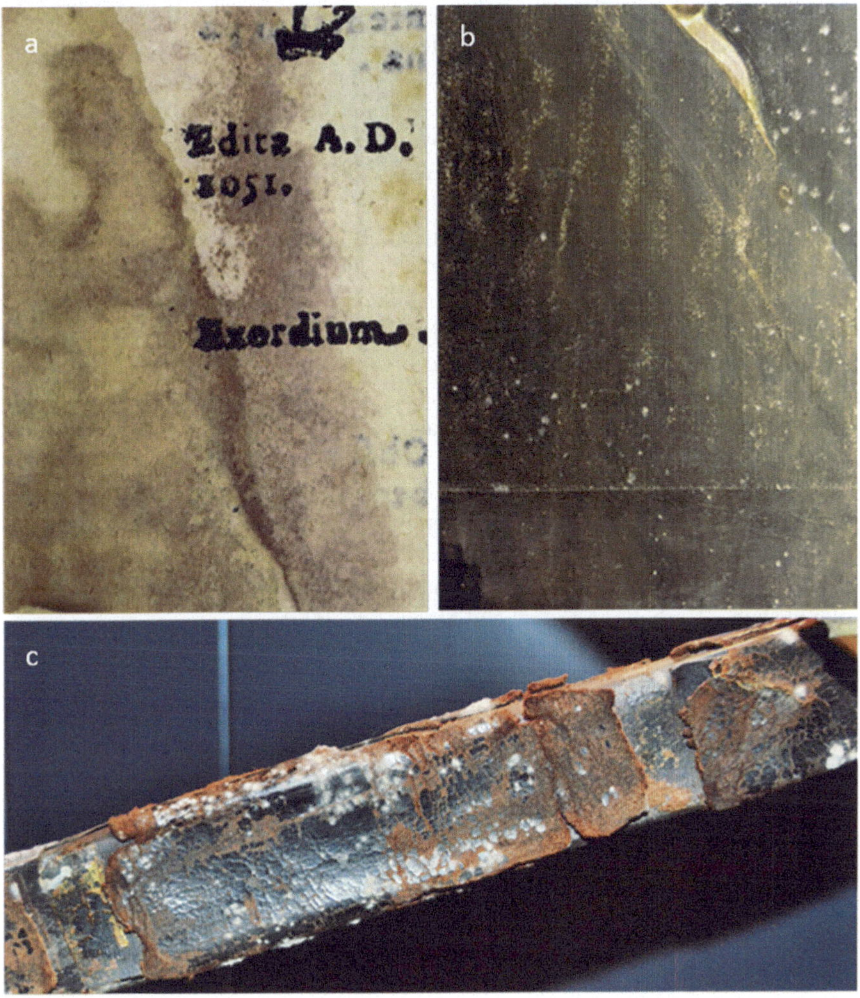

Fig. 1.3 Microbial colonization on organic materials: (**a**) paper document affected of pigmented spots originated by microorganisms, (**b**) an oil painting on canvas with fungal colonization on verso and (**c**) a lather sword holder completely damaged by white fungal colonization

contaminants (López-Miras et al. 2013). Moreover, fungi penetrate cracks and migrate underneath paint layers (Fig. 1.3b), rapidly developing when the environmental relative humidity and temperature increase (RH > 70% and T > 25 °C).

The enzymatic degradation by microorganisms regards also hemicellulose and lignin. The decomposition activity in wood can induce physical-mechanical damage (brown rot, soft rot, and white rot) and chromatic alterations (chromogenic fungi). Chromogenic fungi involve chromatic alterations in wooden objects, producing pigments or metabolic substances. These fungi cause the pinkish, bluish, or grayish discoloration of wood, which may occur either in depth or superficially. Instead, bacteria belonging to *Bacillus*, *Pseudomonas*, *Cellulomonas*, *Clostridium*, and *Cytophaga* genera may induce several alterations in wooden items, with characteristic conformation, i.e., erosion and tunneling (Blanchette 2000).

Textiles, particularly those composed of natural organic fibers, such as cotton, linen, wool, etc., are readily attacked by microorganisms. Microbial growth on textile items causes loss of strength and elongation, discoloration, and changes in appearance (Szostak-Kotowa 2004).

The biodegradation of proteinaceous materials—such as vellum, parchment, and leather, can be induced by bacterial (*Bacillus*, *Pseudomonas*, *Clostridium*, *Streptomycetes*) and fungal (*Mucor*, *Chaetonium*, *Aureobasidium*, *Thrichoderma*, *Epicoccum*) strains. Both groups of microorganisms produce many strong acids and enzymes which can efficiently hydrolyze organic materials, such as extracellular proteolytic enzymes (collagenase, keratinase, etc.). Among actinomycetes (that find proper conditions to proliferate in humid and poorly ventilated environments), there are several strains with strong proteo- and collagenolytic properties, especially within the genus of *Streptomycetes*. Chromatic alterations in leather artifacts, induced by microbial activity, are generally originated by pigmented microbial species (Piñar et al. 2015; Koochakzaei and Achachluei 2015), such as *Streptomyces fimicarius*, generally isolated from miniature and responsible for red- and purple-colored stains (Karbowska-Berent and Strzelczyk 2000).

1.4 Integrated Approach to Reveal and Identify Bacteria and Fungi Colonization on Works of Art Surfaces

The evaluation of those factors promoting microbial activity and proliferation on artworks and the understanding of related deterioration mechanisms is essential in designing appropriate conservation and restoration strategies. Knowledge of microbial contamination of cultural heritage is of great interest, not only in terms of recognizing microbial communities with potential dangerous effects for stored materials or monuments but also in terms of recognizing microorganisms that represent a risk for human health. This justifies the need for performing systematic microbiological sampling to estimate the prevalence of microbial contamination (Borrego et al. 2010); for an evaluation of the biological risk, it is important to measure both the total microbial load in the indoor environment and on artwork surfaces.

Most of the literature reports on studies identifying biodeterioration on cultural heritage using invasive sampling and culture methods. However, traditional culture methods do not always succeed in isolating microbial agents (Michaelsen et al. 2010). Studies using noninvasive sampling and molecular approaches to investigate the role of microorganisms in the deterioration process allow a wider understanding of the biodeterioration phenomena. Generally, a total understanding of the conservation situation can be obtained by an integrated approach using traditional culture techniques, microscopy investigations, and molecular analyses. Instead, the diversity of the cultivable microorganisms adhering to a surface can be analyzed directly by molecular techniques (e.g., RAPD analysis, DGGE fingerprinting). In addition, molecular techniques are increasingly applied in the field of conservation and represent very useful tools for detecting and identifying different microbial strains, but they can never be helpful in the evaluation of damage-causing biodeterioration on the substrate. If combined with microscopic observation, molecular analysis enables interrelations between microorganisms and substrates to be detected (Salvadori 2000).

The study of the mechanisms underlying the microbiological attack of historical materials has been widely practiced and continues to represent one of the main focuses of those institutions and laboratories involved in cultural heritage conservation.

1.4.1 Sampling by Noninvasive Methods

In order to reveal microbial communities and understand their relationship with colonized substrates, the use of noninvasive sampling techniques is often required to minimize invasive actions. Depending on the importance, features, materials, and state of conservation of cultural items, it is not always possible to practice invasive sampling (e.g., use of scalpel to sample small pieces from substrate). This step is also required to monitor microbial colonization before and after conservation and restoration treatments of artistic objects. Several nondestructive methods have been proposed for direct or indirect evidence of microbial colonization or biofilms on different substrates. Current sampling procedures involve sterile needles, adhesive tape, and sterile swabs (Fig. 1.4a, b); the use of these tools is closely related to the nature of sampling objects (Sterflinger 2010; Pasquarella et al. 2015).

In the case of paper items, needles and adhesive tape may cause severe surface damage, so it is preferable to use dry swabs, but unfortunately, these are very often ineffective (Michaelsen et al. 2009). In order to overcome this problem, other noninvasive sampling methods involving the use of nitrocellulose or nylon membrane filters may be adopted and then transferred onto solid culture media to isolate bacteria and fungi (Fig. 1.4c, d), or used to direct DNA extraction (Montanari et al. 2012; Palla 2011).

Nitrocellulose and nylon membranes can be applied to investigate the presence of viable microbial communities that may be the result of aesthetic or structural damage on several materials. In particular, the use of these filters is indicated for ancient paper or leather items (Palla 2006; Pasquarella et al. 2015, 2016).

Microbiologists have generally used sterile adhesive tape to collect samples from skin, food, or medical devices, to detect pathogenic microorganisms in a noninvasive way (Miller 2006); adhesive tape may be associated with microscopic analysis, for example, to detect the presence of fungal conidia on stone surfaces (observation in light microscopy), or with other techniques, as well as fluorescent in situ hybridization (FISH), molecular and culture techniques (La Cono and Urzì 2003; Bisha and Brehm-Stecher 2009). Adhesive tape sampling can be widely and safely applied in the field for the conservation of cultural heritage, due to its several

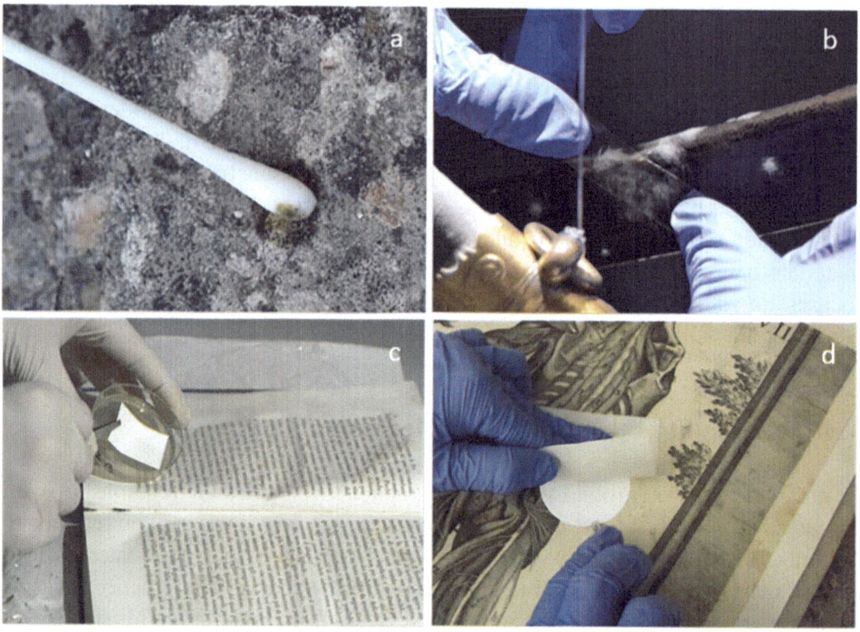

Fig. 1.4 Noninvasive sampling tools: (**a**) cotton swab on stone, (**b**) adhesive tape on lather, (**c**) nylon membrane fragment on book and (**d**) nitrocellulose membrane on paper (drawing)

advantages (i.e., it is cheap and does not require the expertise of specialized staff), but depending on the fragility of surface (Kisová et al. 2020). Urzì and De Leo (2001) proposed the use of adhesive tape strips as a nondestructive sampling technique for the study of microbial colonization on monument surfaces. Cutler et al. (2012) used adhesive tape to collect fungi samples suitable for molecular analysis from stone buildings.

1.4.2 Microscopy

The most common tools used in the field of cultural heritage (assessing the biological contribution to depleting of cultural items) are microscopy techniques applied to the study of biodeterioration processes in cultural assets. Different kinds of microscopes are used to observe biofragments from needle, scalpel, or adhesive tape samplings; to recognize the biodeteriogen morphology, species, secondary metabolites, and products; or to provide direct evidence of biofilm formation.

Microscopy techniques are also widely used to observe the morphology of microbial taxa; several microscopy techniques have been used to study the relationship between microorganisms and rock applying nondestructive methods; both light microscopy and light microscopy combined with scanning electron microscopy were used (de los Rìos and Ascaso 2005).

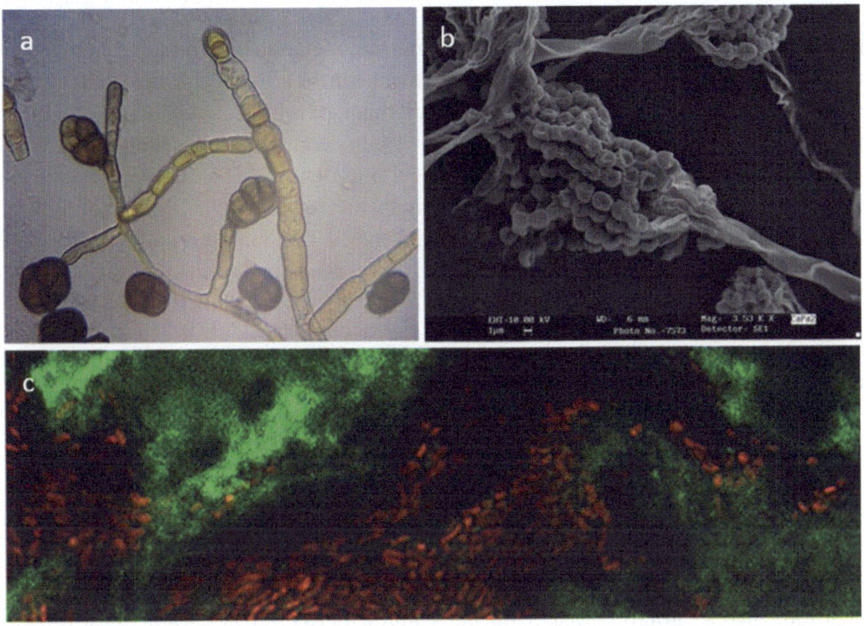

Fig. 1.5 Microscopy techniques used to reveal microbial systems: (**a**) fungal structures (*Alternaria* spp.) stained by Lugol's solution observed by OM—40× magnification, (**b**) micrograph by SEM of fungal reproductive structure and spores (*Penicillium* spp.), and (**c**) *green and red autofluorescence patterns* related to cyanobacteria and algae cells by CLSM

The most common microscopy techniques used in studying microbial colonies are optical microscopy (OM), or light microscopy (LM), scanning electron microscopy with back-scattered electron imaging (SEM-BSE), transmission electron microscopy (TEM), confocal laser scanning microscopy (CLSM), and environmental SEM (ESEM) (Macedo et al. 2009; Herrera and Videla 2009; Rakotonirainy et al. 2007).

Optical or light microscopy can provide information simply by observing the samples and is useful in the detection of several morphological structures (Fig. 1.5a) (e.g., fungi, cyanobacteria, algae, lichens) (Golubíc et al. 2015; Macedo et al. 2009). This type of microscope is useful for first stage identification of bacteria as cellular morphology and reaction with the Gram stain can be verified (Bergmans et al. 2005).

Electronic microscopes are widely used in scientific applications, the most frequently utilized being SEM and TEM, which allow observation and surface characterization at a higher resolution and depth of field than the optical microscope by giving a 3D effect to the images (Fig. 1.5b) (Nuhoglu et al. 2006; Troiano et al. 2014; Golubíc et al. 2015; Florian and Manning 2000; Cappitelli et al. 2009). Unlike SEM, in ESEM analysis the sample is hydrated and nonconducting, without prior dehydration or conductive coating (Bergmans et al. 2005).

CLSM is extremely useful for revealing the chemical and biological relationship between a microorganism (biodeteriogen) and its microenvironment (colonized substrate); CLSM has many advantages, as its nondestructive and "real-time" technique can be used in situ to detect the growth, metabolism, and gene expression of organisms including the microenvironment surrounding each cell (Fig. 1.5c) (Caldwell et al. 1992; Villa et al. 2015). A full range of multi-microscopic techniques (CLSM/SEM) is often used to detect biota communities (fungi, bacteria, cyanobacteria) on samples from particular cultural items (e.g., mummies, fossilized bones), to confirm the presence of certain organisms, or to establish which dominant biota is causing deterioration in biofilm (Palla et al. 2011; Janssens and Van Grieken 2004; Marano et al. 2016; Polo et al. 2012; de los Rìos and Ascaso 2005).

1.4.3 In Vitro Culture

Traditional culture methods for isolation and identification of microorganisms are very useful in assessing biodeterioration in cultural heritage. The cultivation methods (nutritive media) are used for the specific isolation of fungi and bacteria or specific growing of other biological systems, such as cyanobacteria or lichens. Some nutritive media may be designed for specific strains proposing the growth condition that originally existed on the cultural heritage (e.g., high pH level for strains isolated from wall paintings) (Gorbushina et al. 2004).

In order to reveal microbial contamination of cultural items (organic or inorganic), it is possible to place a piece of item (sample) directly onto the nutritive medium (Abrusci et al. 2005) or sampling tools, such as swabs, adhesive tape, or charged membranes (Pasquarella et al. 2015; Palla et al. 2015). No medium has been recommended for the isolation of certain microorganisms; as regards the isolation of fungal communities from paper heritage, for example, Sabouraud dextrose agar and malt extract agar, sometimes supplemented with an antibiotic such as chloramphenicol or streptomycin to inhibit bacterial growth, and Czapek agar have been used. For determining the total number of cultivable bacteria, tryptic soy agar is used, while MacConkey agar is used to detect gram-negative bacteria (López-Miras et al. 2013).

The microbial samples may be put into a nutritive broth (generally composed of a peptone solution with meat extract and mineral salts, agar-free) and incubated for many hours (18–72 h) to allow microbial growth.

Thus, it is possible to perform morphological analysis detecting and evaluating macroscopic parameters, e.g., color, shape, and appearance (Fig. 1.6). Colony-forming units (CFU) on solid growth media are counted using a simple visual inspection. Besides pH, the availability of nutrients, the temperature, and the duration of incubation are also important for successful isolation. However, methods based on the cultivation of microorganisms may detect only a minor fraction (2–5%) of the total number of microbial communities, this cultivable fraction representing only the living part of the microbes in the sample (Otlewska et al. 2014). In any case, culture-based approaches alone cannot provide exhaustive information on the real microbial consortia, also because only a small fraction of the microorganisms can be cultivated (Dakal and Arora 2012). For this reason, for the correct evaluation of

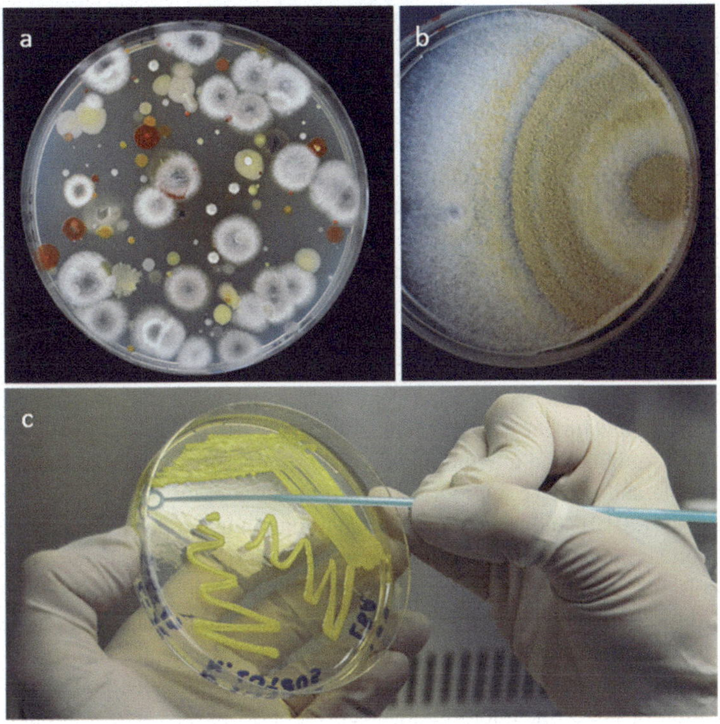

Fig. 1.6 Bacteria and fungi isolated by culture-dependent methods: (**a**) huge microbial growing on nutritive medium, (**b**) fungal colony on Sabouraud media (*Scopulariopsis* spp.), and (**c**) subculture onto N-Agar medium (*Micrococcus luteus*)

biological contaminants, culture techniques should be supported by molecular investigations.

1.4.4 Molecular Investigations

While traditional microbiological methods based on culture procedures provide important but limited information on microbial communities, molecular techniques offer a great overview on the diversity of microbiota involved in the biodeterioration of monuments and artworks (Otlewska et al. 2014; Gonzáles and Saiz-Jiménez 2005).

The development of recent molecular techniques has improved the sensitivity, specificity, and speed of detecting microorganisms. Primarily, the molecular detection of microorganisms begins with the extraction of nucleic acids from collected samples, using specific ad hoc protocols or commercial kits. The detection of microorganisms is mainly based on the sequence of small subunits (16S for bacteria and 18S for fungi) ribosomal RNA (rRNA) genes (Gonzáles and Saiz-Jiménez

2005). Molecular investigation continues with in vitro amplification of specific genomic regions by polymerase chain reaction (PCR), analyses of amplification products using gel electrophoresis technique, determination of nucleotide composition of DNA fragments (sequencing) and analyses of homology of sequence using international data banks (e.g., NIH, USA; EMBL, Germany). The existence of DNA database guarantees optimal identification of the microorganisms detected and the possibility of carrying out phylogenetic analysis (Palla 2004; Gonzáles and Saiz-Jiménez 2005).

At present, molecular methods, including polymerase chain reaction (PCR), denaturing gradient gel electrophoresis (DGGE) and the creation of clone libraries, are used as a sensitive alternative to conventional cultivation techniques. DGGE is the most frequently reported technique for separating DNA fragments during microbial diversity studies of art objects (Dakal and Arora 2012; Otlewska et al. 2014; González and Saiz-Jiménez 2004; Michaelsen et al. 2010). Genetic fingerprinting is a rapid and useful method for studying diversity in microbial communities including nonculturable and inactive microorganisms (Otlewska et al. 2014).

Polo et al. (2010) studied microbial contamination using denaturing gradient gel electrophoresis and established that cyanobacteria and green algae genera were responsible for green staining. Fluorescence in situ hybridization (FISH) has been applied in the field of conservation and restoration to study microbiota involved in biodeterioration. Furthermore, the application of FISH directly on adhesive tape strips added another advantage to this nondestructive sampling method and led to the identification in situ of the microorganisms present on a given area, without destruction of the valuable surfaces and with little biofilm disturbance (Andersen et al. 2001; Sterflinger 2010).

The application of molecular biology on cultural heritage has been applied to identify the development of microorganisms in inorganic materials or in organic materials, in museums, and in archives (Piñar et al. 2013; Valentin 2003; Palla 2020). Molecular biology provides a sensitive study on the microbial contamination of works of art, based on the analysis of specific DNA genomic sequences.

In order to increase the screening of a large number of microorganisms, oligonucleotide microarray protocols have been applied for the detection and genotyping microbial variety (Maynard et al. 2005) and can be used efficiently in the identification of microbial communities affecting cultural heritage.

Microarray analysis is based on hybridization between fluorescent-labeled DNA sequences and thousands of probes immobilized on a glass slide (array), providing a cultivation-independent characterization of microbes on works of art (Gargano et al. 2012; Neugebauer et al. 2010). The core principle behind microarrays is hybridization between two DNA strands, the property of complementary nucleic acid sequences to specifically pair with each other by forming hydrogen bonds between complementary nucleotide base pairs. Each DNA spot contains a specific DNA probe sequence, used to hybridize a sample called target under high-stringency conditions. Fluorescently labeled target sequences that bind to a probe sequence generate a signal that depends on hybridization conditions (such as temperature) and washing after hybridization. These sample microarrays contain a large number of samples that could be analyzed, for instance, through FISH techniques for the detection of specific microorganisms or the expression of specific genes (Gonzalez 2003; Andersen et al. 2001; Kononen et al. 1998).

References

Abrusci C, Martìn-Gonzalez A, Del Amo A, Catalina F, Collado J, Platas G (2005) Isolation and identification of bacteria and fungi from cinematographic films. Int Biodeterior Biodegradation 56(1):58–68. https://doi.org/10.1016/j.ibiod.2005.05.004

Adamo P, Violante P (2000) Weathering of rocks and neogenesis of minerals associated with lichen activity. Appl Clay Sci 16(5–6):229–256. https://doi.org/10.1016/S0169-1317(99)00056-3

Albertano P, Urzì C (1993) Structural interactions among epilithic cyanobacteria and heterotrophic microorganisms in Roman hypogea. Microbial Ecol 38(3):244–252. https://doi.org/10.1007/s002489900174

Albertano P, Moscone D, Palleschi G, Hermosin B, Saiz-Jimenez C, Sanchez-Moral S, Hernandez-Marine M, Urzì C, Groth I, Schroeckh V, Saarela M, Mattila-Sandholm T, Gallon JR, Graziottin F, Bisconti F, Giuliani R (2003) Cyanobacteria attack rocks (CATS): control and preventive strategies to avoid damage caused by cyanobacteria and associated microorganisms in Roman hypogean monuments. In: Saiz-Jimenez C (ed) Molecular biology and cultural heritage. Balkema, Lisse

Allsopp D, Seal KJ, Gaylarde CC (2004) Introduction to biodeterioration. Cambridge University Press, Cambridge

Almeida MT, Mouga T, Barracosa P (1994) The weathering ability of higher plants. The case of Alianthus altissima (Miller) Swingle. Int Biodeterior Biodegradation 33(4):333–343. https://doi.org/10.1016/0964-8305(94)90011-6

Anagnostidis K, Gehrmann CK, Gross M, Krumbein WE, Lisi S, Pantazidou A, Urzì C, Zagari M (1991) Biodeterioration of marbles of the Parthenon and propylaea, acropolis, Athens-associated organisms, decay and treatment suggestions. In: Decrouez D, Chamay J, Zezza F (eds) Proceedings of the 2nd international symposium. Musée dArt et dHistoire de Genéve, Genéve, pp 305–325

Andersen CL, Hostetter G, Grigoryan A, Sauter G, Kallioniemi A (2001) Improved procedure for fluorescence in situ hybridization on tissue microarrays. Cytometry 45:83–86

Ascaso C, Wierzchos J (1995) Study of the biodeterioration zone between the lichen thallus and the substrate. Cryptogam Bot 5:270–281

Barresi G, Di Carlo E, Trapani MR, Parisi MG, Chillè C, Mulè MF, Cammarata M, Palla F (2015) Marine organism as source of bioactive molecules applied in restoration projects. Herit Sci 3(17). https://doi.org/10.1186/s40494-015-0046-1

Bergmans L, Moisiadis P, Van Meerbeek B, Quirynen M, Lambrechts P (2005) Microscopic observation of bacteria: review highlighting the use of environmental SEM. Int Endod J 38(11):775–788. https://doi.org/10.1111/j.1365-2591.2005.00999.x

Bettini C, Villa A (1976) Il problema della vegetazione infestante nelle aree archeologiche. In: Rossi-Manaresi R (ed) The conservation of stone I, Bologna, pp 191–204

Bisha B, Brehm-Stecher BF (2009) Simple adhesive-tape-based sampling of tomato surfaces combined with rapid fluorescence in situ hybridization for salmonella detection. Appl Environ Microbiol 75(5):1450–1455. https://doi.org/10.1128/AEM.01944-08

Biswas J, Sharma K, Harris KK, Rajput Y (2013) Biodeterioration agents: bacterial and fungal diversity dwelling in or on the pre-historic rock-paints of Kabra-pahad. India Iran J Microbiol 5(3):309–314

Blanchette RA (2000) A review of microbial deterioration found in archaeological wood from different environments. Int Biodeterior Biodegradation 46:189–204. https://doi.org/10.1016/S0964-8305(00)00077-9

Bock E, Koops HP, Moller UC, Rudert M (1990) A new facultatively nitrite oxidizing bacterium, Nitrobacter vulgaris sp nov. Arch Microbiol 153:105–110
Borrego S, Guiamet P, Gómez de Saravia S, Batistini P, Garcia M, Lavin P, Perdomo I (2010) The quality of air at archives and the biodeterioration of photographs. Int Biodeterior Biodegradation 64(2):139–145. https://doi.org/10.1016/j.ibiod.2009.12.005
Borrego S, Lavin P, Perdomo I, Gómez de Saravia S, Guiamet P (2012) Determination of indoor air quality in archives and biodeterioration of the documentary heritage. ISRN Microbiol 2012:1–10. https://doi.org/10.5402/2012/680598
Burford EP, Fomina M, Gadd GM (2003) Fungal involvement in bioweathering and biotransformation of rocks and minerals. Mineral Mag 67:1127–1155
Caldwell DE, Korber DR, Lawrenc JR (1992) Confocal laser microscopy and digital image analysis in microbial ecology. Adv Microb Ecol 12:1–67
Caneva G, Galotta G (1994) Floristic and structural changes of plant communities of the Domus Aurea (Rome) related to a different weed control. In: Fassina V, Off H, Zezza F (eds) Proceedings of the 3rd international symposium. The conservation of monuments in the mediterranean basin. ICCROM, Venezia, pp 317–322
Caneva G, Nugari MP, Salvadori O (1994) La biologia nel restauro. Nardini Editore, Firenze
Caneva G, Nugari MP, Salvadori O, Getty Conservation Institute (1998) Plant biology for cultural heritage: biodeterioration and conservation. Getty Publications, Los Angeles
Caneva G, Pacini A, Celesti Grapow L, Ceschin S (2003) The Colosseum's use and state of abandonment as analysed through its flora. Int Biodeterior Biodegradation 51(3):211–219. https://doi.org/10.1016/S0964-8305(02)00173-7
Caneva G, Ceschin S, De Marco G (2006) Mapping "tree root damage risk" for the conservation of archaeological sites: the case of the Domus Aurea (Rome). Conserv Manage Archaeol Sites 7(3):163–170
Caneva G, Nugari MP, Salvadori O (2008) Plant biology for cultural heritage. The Getty Conservation Institute. Biodeterioration and conservation, Los Angeles
Caneva G, De Nuntiis P, Fornaciari M, Ruga L, Valenti P, Pasquariello G (2020) Aerobiology applied to the preventive conservation of cultural heritage. Aerobiologia 36:99–103. https://doi.org/10.1007/s10453-019-09589-9
Canhoto O, Pinzari F, Fanelli C, Magan N (2004) Application of electronic nose technology for the detection of fungal contamination in library paper. Int Biodeterior Biodegradation 54(4): 303–309. https://doi.org/10.1016/j.ibiod.2004.04.001
Capotorti G, Del Vico E, Lattanzi E, Tilia A, Celesti Grapow L (2013) Exploring biodiversity in a metropolitan area in the Mediterranean region: the urban and suburban flora of Rome (Italy). Plant Biosyst 147(1):174–185. https://doi.org/10.1080/11263504.2013.771715
Cappitelli F, Abbruscato P, Foladori P, Zanardini E, Ranalli G, Principi P, Villa F, Polo A, Sorlini C (2009) Detection and elimination of cyanobacteria from frescoes: the case of the St. Brizio Chapel (Orvieto Chatedral, Italy). Microb Ecol 57:633–639. https://doi.org/10.1007/s00248-008-9441-4
Casoli E, Ricci S, Belluscio A, Gravina MF, Ardizzone G (2015) Settlement and colonization of epi-endobenthic communities on calcareous substrata in an underwater archaeological site. Mar Ecol 36(4):1060–1074
Cataldo R, De Donno A, De Nunzio G, Leucci G, Nuzzo L, Siviero S (2005) Integrated methods for analysis of cultural heritage: the crypt of "Cattedrale di Otranto". J Cult Herit 6(1):29–38. https://doi.org/10.1016/j.culher.2004.05.004
Celesti-Grapow L, Blasi C (2004) The role of Alien and Native Weeds in the deterioration of archaeological remains in Italy. Weed Technol 18:1508–1513
Chen J, Blume HP, Beyer L (2000) Weathering of rocks induced by lichen colonization – a review. Catena 39(2):121–146. https://doi.org/10.1016/S0341-8162(99)00085-5
Crispim CA, Gaylarde CC (2005) Cyanobacteria and biodeterioration of cultural heritage: a review. Microb Ecol 49(1):1–9. https://doi.org/10.1007/s00248-003-1052-5

Cutler NA, Oliver AE, Viles HA, Whiteley AS (2012) Non-destructive sampling of rock-dwelling microbial communities using sterile adhesive tape. J Microbiol Meth 91(3):391–398. https://doi.org/10.1016/j.mimet.2012.09.022

Cutler NA, Viles HA, Ahmad S, McCabe S, Smith BJ (2013) Algal "greening" and the conservation of stone heritage structures. Sci Total Environ 442:152–164. https://doi.org/10.1016/j.scitotenv.2012.10.050

Dakal TC, Arora KP (2012) Evaluation of potential molecular and physical techniques in studying biodeterioration. Rev Environ Sci Biotechnol 11:71–104. https://doi.org/10.1007/s11157-012-9264-0

Dakal TC, Cameotra SS (2011) Geomicrobiology of cultural monuments and artworks: mechanism of biodeterioration, bioconservation strategies and applied molecular approaches. In: Mason AC (ed) Bioremediation: biotechnology, engineering, and environment management. Nova Science Publishers, New York

Dakal TC, Cameotra SS (2012) Microbially induced deterioration of architectural heritages: routes and mechanisms involved. Environ Sci Eur 24(1):1–13. https://doi.org/10.1186/2190-4715-24-36

de la Torre MA, Gomez-Alarcon G (1994) Manganese and iron oxidation by fungi isolated from building stone. Microb Ecol 27(2):177–188

de la Torre MA, Gomez-Alarcon G, Vizcaino C, Garcia MT (1993) Biochemical mechanisms of stone alteration carried out by filamentous fungi living in monuments. Biogeochemistry 19(3): 129–147

de los Rìos A, Ascaso C (2005) Contribution of in situ microscopy to the current understanding of stone biodeterioration. Int Microbiol 8(3):181–188

de los Ríos A, Cámara B, García del Cura MÁ, Rico VJ, Gálvan V, Ascaso C (2009) Deteriorating effects of lichen and microbial colonization of carbonate building rocks in Romanesque churches of Segovia (Spain). Sci Total Environ 407(3):123–1134. https://doi.org/10.1016/j.scitotenv.2008.09.042

Delgado Rodrigues J, Valero J (2003) A brief note on the elimination of dark stains of biological origin. Stud Conserv 48(1):17–22

Fernandes P (2006) Applied microbiology and biotechnology in the conservation of cultural heritage materials. Appl Microbiol Biotechnol 73(2):291–296. https://doi.org/10.1007/s00253-006-0599-8

Florian MLE, Manning L (2000) SEM analysis of irregular fungal fox spots in an 1854 book: population dynamics and species identification. Int Biodeterior Biodegradation 46(3):205–220. https://doi.org/10.1016/S0964-8305(00)00062-7

Gargano V, Mancuso FP, Vitale F, Reale S, Caracappa S, Palla F (2012) DNA-microarray technology for the identification of microbial species onto the surfaces and in the aerosol of indoor environments. In: Sistemi biologici e Beni Culturali: area tematica Biologia e Biotecnologie per i Beni Culturali, atti Convegno Nazionale AIAr, Palermo, 6–7 Ottobre 2009, Regione Siciliana, Assessorato dei Beni Culturali e dell'Identità Siciliana, Dipartimento dei Beni Culturali e dell'Identità Siciliana, pp 164–171

Giustetto R, Gonella D, Bianciotto V, Lumini E, Voyron S, Costa E, Diana E (2015) Transfiguring biodegradation of frescoes in the Beata Vergine del Pilone sanctuary (Italy): microbial analysis and minero-chemical aspects. Int Biodeterior Biodegradation 98:6–18. https://doi.org/10.1016/j.ibiod.2014.10.020

Golubíc S, Pietrini AM, Ricci S (2015) Euendolithic activity of cyanobacterium Chroococcus lithophilus Erc. In biodeterioration of the pyramid of Caius Cestius, Rome, Italy. Int Biodet Biodeg 100:7–16. https://doi.org/10.1016/j.ibiod.2015.01.019

Gonzáles JM, Saiz-Jiménez C (2005) Application of molecular nucleic acid-based techniques for the study of microbial communities in monuments and artworks. Int Microbiol 8(3):189–194

Gonzalez JM (2003) Overview on existing molecular techniques with potential interest in cultural heritage. In: Taylor & Francis (ed) Molecular biology and cultural heritage. Balkema, Lisse, pp 3–13

González JM, Saiz-Jiménez C (2004) Microbial diversity in biodeteriorated monuments as studied by denaturing gradient gel electrophoresis. J Sep Sci 27(3):174–180. https://doi.org/10.1002/jssc.200301609

Gorbushina AA, Palinska KA (1999) Biodeteriorative processes on glass: experimental proof of the role of fungi and cyanobacteria. Aerobiologia 15(3):183–191. https://doi.org/10.1023/A:1007616614172

Gorbushina AA, Heyrman J, Dornieden T, Gonzalez-Delvalle M, Krumbein WE, Laiz L, Petersen K, Saiz-Jimenez C, Swings J (2004) Bacterial and fungal diversity and biodeterioration problems in mural paintings environments of St. Martins Church (Greene-Kreiensen, Germany). Int Biodeter Biodegr 53(1):13–24. https://doi.org/10.1016/j.ibiod.2003.07.003

Green CF, Scarpino PV (2002) The use of ultraviolet germicidal irradiation (UVGI) in disinfection of airborne bacteria. Environ Eng Policy 3(1):101–107. https://doi.org/10.1007/s100220100046

Gu JD, Ford TE, Berke NS, Mitchell R (1998) Biodeterioration of concrete by the fungus fusarium. Int Biodeterior Biodegradation 41(1):101–109. https://doi.org/10.1016/S0964-8305(98)00034-1

Guiamet P, Igareta A, Battistoni P, Gómez de Saravia S (2014) Fungi and bacteria in the biodeterioration of archeological fibers. Analysis using different microscopic techniques. Rev Argent Microbiol 46(4):376–377. https://doi.org/10.1016/S0325-7541(14)70097-5

Gutarowska B (2010) Metabolic activity of moulds as a factor of building materials biodegradation. Pol J Microbiol 59(2):119–124

Hauer T, Mühlsteinová R, Bohunická M, Kaštovský J, Mareš J (2015) Diversity of cyanobacteria on rock surfaces. Biodivers Conserv 24(4):759–779. https://doi.org/10.1007/s10531-015-0890-z

Herrera LK, Videla HA (2009) Surface analysis and materials characterization for the study of biodeterioration and weathering effects on cultural property. Int Biodeterior Biodegradation 63(7):813–822. https://doi.org/10.1016/j.ibiod.2009.05.002

Herrera LK, Le Borgne S, Videla HA (2009) Modern methods for materials characterization and surface analysis to study the effects of biodeterioration and weathering on buildings of buildings of cultural heritage. Int J Archit Herit 3(1):74–91. https://doi.org/10.1080/15583050802149995

Hirsch P, Eckhardt FEW, Palmer RJ (1995) Methods for the study of rock-inhabiting microorganisms-a mini review. J Microbiol Methods 23(2):143–167. https://doi.org/10.1016/0167-7012(95)00017-F

Hochella MF (2002) Sustaining earth: thoughts on the present and future roles in mineralogy in environmental science. Mineral Mag 66(5):627–652. https://doi.org/10.1180/0026461026650053

Hoffmann L (2002) Caves and other low-light environments: aerophytic photoautotrophic microorganisms. In: Bitton G (ed) Encyclopedia of environmental microbiology. Wiley, New York, pp 835–843

Hoppert M, König S (2006) The succession of biofilms on building stone and its possible impact on biogenic weathering. In: Fort R, Alvarez de Buergo M, Gomez-Heras M, Vazquez-Calvo C (eds) Heritage, weathering and conservation, vol 2. Taylor & Francis, London, pp 311–315

Hu H, Ding S, Katayama Y, Kusumi A, Xian Li S, de Vries RP, Wang J, Yu X-Z, Gu J-D (2013) Occurrence of *aspergillus allahabadii* on sandstone at Bayon temple, Angkor Thom, Cambodia. Int Biodeterior Biodegradation 76:112–117. https://doi.org/10.1016/j.ibiod.2012.06.022

Hueck HJ (1968) The biodeterioration of materials – an appraisal. In: Walters AH, Elphick JS (eds) Biodeterioration of materials. Elsevier, London, pp 6–12

Huer B (2008) Spatial distribution of cyanobacteria and algae from the tombstone in a historic cemetery in Bratislava. Slovakia Fottea 9(1):81–92

Isola D, Zucconi L, Onofri S, Caneva G, Hoog GS, Selbmann L (2016) Extremotolerant rock inhabiting black fungi from Italian monumental sites. Fungal Divers 76(1):1–22. https://doi.org/10.1007/s13225-015-0342-9

Janssens K, Van Grieken R (eds) (2004) Non-destructive micro analysis of cultural heritage materials, vol 42. Elsevier, Amsterdam

Karbowska-Berent J, Strzelczyk A (2000) The role of *Streptomycetes* in the biodeterioration of historic parchment. Acta Microbiol Pol 49:177–178
Karbowska-Berent J, Jarmiłko J, Czuczko J (2014) Fungi in fox spots of a drawing by Leon Wyczółkowski. Restaurator 35(2):159–179. https://doi.org/10.1515/rest-2014-1000
Kauffmann MJ (1952) Rôle des bactéries nitrificantes dans l'altération des pierres calcaires des monuments. C R Acad Sci 234:2395–2397
Kavkler K, Gunde-Cimerman N, Zalar P, Demšar A (2015) Fungal contamination of textile objects preserved in Slovene museums and religious institutions. Int Biodeterior Biodegradation 97:51–59. https://doi.org/10.1016/j.ibiod.2014.09.020
Kisová Z, Planý M, Pavlović J, Bučková M, Puškárová A, Kraková L, Kapustová M, Pangallo D, Šoltys K (2020) Biodeteriogens characterization and molecular analyses of diverse funeral accessories from XVII Century. Appl Sci 10(16):5451. https://doi.org/10.3390/app10165451
Kononen J, Bubendorf L, Kallioniemi A, Barlund M, Schraml P, Leighton S, Torhorst J, Mihatsch MJ, Sauter G, Kallioniemi OP (1998) Tissue microarrays for high-throughput molecular profiling of tumor specimens. Nat Med 4(7):844–847. https://doi.org/10.1038/nm0798-844
Koochakzaei A, Achachluei MM (2015) Red stains on archaeological leather: degradation characteristics of a shoe from the 11th-13th centuries (Seljuk period, Iran). JAIC 54(1):45–56. https://doi.org/10.1179/1945233014Y.0000000033
Kraková L, Chovanová K, Selim SA, Šimonovičová A, Puškarová A, Maková A, Pangallo D (2012) A multiphasic approach for investigation of microbial diversity and its biodegradative abilities in historical paper and parchment documents. Int Biodeterior Biodegradation 70:117–125. https://doi.org/10.1016/j.ibiod.2012.01.011
Krumbein WE (2003) Patina and cultural heritage – a geomicrobiologist's prospective – how microbes change surfaces. In: Kozlowski R (ed) Cultural Heritage Research: a Pan European Challenge, Proceedings of the 5th European Commission Conference, Cracow, Poland, 16–18 May 2002, pp 1–9
Kumar R, Kumar AV (1999) Biodeterioration of stone in tropical environments. An overview, The Getty Conservation Institute, Los Angeles
La Cono V, Urzì C (2003) Fluorescent in situ hybridation applied on samples taken with adhesive tape strips. J Microbiol Methods 55(1):65–71. https://doi.org/10.1016/S0167-7012(03)00115-5
La Placa M (2005) Principi di microbiologia medica. Esculapio Editore, Bologna
Lamenti G, Tomaselli L, Tiano P (2003) Cyanobacteria and biodeterioration of monumental stones. In: Saiz-Jimenez C (ed) Molecular biology and cultural heritage. Balkema, Lisse, pp 73–78
Lisci M, Monte M, Pacini E (2003) Lichens and higher plants on stone: a review. Int Biodeterior Biodegradation 51(1):1–17. https://doi.org/10.1016/S0964-8305(02)00071-9
Lombardozzi V, Castrignanò T, D'Antonio M, Casanova Municchia A, Caneva G (2012) An interactive database for an ecological analysis of stone biopitting. Int Biodeterior Biodegradation 73:8–15. https://doi.org/10.1016/j.ibiod.2012.04.016
López-Miras M, Piñar G, Romero-Noguera J, Bolivar-Galiano FC, Ettenauer J, Sterflinger K, Martin-Sanchez I (2013) Microbial communities adhering to the obverse and reverse sides of an oil painting on canvas: identification and evaluation of their biodegradative potential. Aerobiologia 29(2):301–314. https://doi.org/10.1007/s10453-012-9281-z
Macedo MF, Miller AZ, Dionísio A, Saiz-Jimenez C (2009) Biodiversity of cyanobacteria and green algae on monuments in the Mediterranean Basin: an overview. Microbiology 155(11): 3476–3490. https://doi.org/10.1099/mic.0.032508-0
Mandrioli P, Caneva G, Sabbioni C (eds) (2003) Cultural heritage and aerobiology: methods and measurement techniques for biodeterioration monitoring. Springer, Cham. https://doi.org/10.1007/978-94-017-0185-3
Manente S, Micheluz A, Ganzerla R, Ravagnan G, Gambaro A (2012) Chemical and biological characterization of paper: a case study using a proposed methodological approach. Int Biodeterior Biodegradation 72:99–108. https://doi.org/10.1016/j.ibiod.2012.03.008

Mansch R, Bock E (1998) Biodeterioration of natural stone with special reference to nitrifying bacteria. Biodeterioration 9(1):47–64. https://doi.org/10.1023/A:1008381525192
Marano F, Di Rita F, Palombo MR, Ellwood NTW, Bruno L (2016) A first report of biodeterioration caused by cyanobacterial biofilms of exposed fossil bones: a case study of the middle Pleistocene site of La Polledrara di Cecanibbio (Rome, Italy). Int Biodeterior Biodegradation 106:67–74. https://doi.org/10.1016/j.ibiod.2015.10.004
Martino M, Schiavone S, Balloi A, Pellegrino L, De Castro E, Palla F (2015) Bioremoval of sulphate layer from a 15th century polychrome marble artefact. Cons Sci Cult Herit 15:235–243
Marvasi M, Donnarumma F, Frandi A, Mastromei G, Sterflinger K, Tiano P, Perito B (2012) Black microcolonial fungi as deteriogens of two famous marble statues. Int Biodeterior Biodegradation 68:36–44. https://doi.org/10.1016/j.ibiod.2011.10.011
May E, Lewis FJ, Pereire S, Tayler S, MRD S, Allopp D (1993) Microbial deterioration of building stone – a review. Biodeterioration Abstracts 7:109–123
Maynard C, Berthiaume F, Lemarchand K, Harel J, Payment P, Bayardelle P, Masson L, Brousseau R (2005) Waterborne pathogens detection by use of oligonucleotide-based, microarray. Appl Environ Microbiol 71(12):8548–8557. https://doi.org/10.1128/AEM.71.12.8548-8557.2005
McNamara CJ, Mitchell R (2005) Microbial deterioration of historic stone. Front Ecol Environ 3(8): 445–451. https://doi.org/10.1890/1540-9295(2005)003[0445:MDOHS]2.0.CO;2
Michaelsen A, Piñar G, Montanari M, Pinzari F (2009) Biodeterioration and restoration of a 16th century book using a combination of conventional and molecular techniques: a case study. Int Biodeterior Biodegradation 63(2):161–168. https://doi.org/10.1016/j.ibiod.2008.08.007
Michaelsen A, Piñar G, Pinzari F (2010) Molecular and microscopical investigation of the microflora inhabiting a deteriorated Italian manuscript dated from the thirteenth century. Microb Ecol 60(1):69–80. https://doi.org/10.1007/s00248-010-9667-9
Miller DL (2006) Application of adhesive techniques to harvest stratum corneum material. In: Serup J, Jemec GBE, Grove GL (eds) Handbook of non-invasive methods and the skin. CRC Press, Second Edition, pp 371–373. https://doi.org/10.3109/9781420003307-52
Miller JD, Haisley PD, Reinhardt JH (2000) Air sampling result in relation to extend of fungal colonization of building materials in some water-damaged buildings. Indoor Air 10(3):146–151. https://doi.org/10.1034/j.1600-0668.2000.010003146.x
Miller A, Macedo MF, Fort R, Alvarez de Buergo M, Gomez-Heras M, Vazquez-Calvo C (2006) Mapping and characterization of a green biofilm inside of Vilar de Frades church (Portugal). In: Heritage, weathering and conservation, vol 2. Taylor & Francis, London, pp 329–355
Mishra AK, Jain KK, Garg KL (1995) Role of higher plants in the deterioration of historical buildings. Sci Total Environ 167(1–3):375–392. https://doi.org/10.1016/0048-9697(95)04597-T
Mitchell R, Gu J-D (2000) Changes in the biofilm microflora of limestone caused by atmospheric pollutants. Int Biodeterior Biodegradation 46(4):299–303. https://doi.org/10.1016/S0964-8305(00)00105-0
Montanari M, Melloni V, Pinzari F (2012) Fungal biodeterioration of historical materials stored in Compactus movable shelves. Int Biodeterior Biodegradation 75:83–88. https://doi.org/10.1016/j.ibiod.2012.03.011
Montemartini Corte A, Ferroni A, Salvo VS (2003) Isolation of fungal species from test samples and maps damaged by foxing, and correlation between these species and the environment. Int Biodeterior Biodegradation 51(3):167–173. https://doi.org/10.1016/S0964-8305(02)00137-3
Moroni B, Pitzurra L (2008) Biodegradation of atmospheric pollutants by fungi: a crucial point in the corrosion of carbonate building stone. Int Biodeterior Biodegradation 62(4):391–396. https://doi.org/10.1016/j.ibiod.2008.03.011
Neelakanta G, Sultana H (2013) The use of metagenomic approaches to analyze changes in microbial communities. Microbiol Insights 6:37–48. https://doi.org/10.4137/MBI.S10819
Neugebauer W, Leinberger DM, Petersen K, Schumacher U, Bachmann TT, Krekel C (2010) The development of a DNA microarray for the rapid identification of moulds on works of art. Stud Conser 55(4):258–273. https://doi.org/10.1179/sic.2010.55.4.258

Nielsen KF, Holm G, Uttrup LP, Nielsen PA (2004) Mould growth on building materials under low water activities. Influence of humidity and temperature on fungal growth and secondary metabolism. Int Biodeterior Biodegradation 54(4):325–336. https://doi.org/10.1016/j.ibiod.2004.05.002

Nittérus M (2000) Fungi in archives and libraries, a literary survey. Restaurator 21(1):25–40. https://doi.org/10.1515/REST.2000.25

Nuhoglu Y, Oguz E, Uslu H, Ozbek A, Ipekoglu B, Ocak I, Hasenekoglu I (2006) The accelerating effects of the microorganisms on biodeterioration of stone monuments under air pollution and continental-cold climatic conditions in Erzurum. Turkey Sci Total Environ 364(1–3):272–283. https://doi.org/10.1016/j.scitotenv.2005.06.034

Onofri S, Zucconi L, Isola D, Selbmann L (2014) Rock-inhabiting fungi and their role in deterioration of stone monuments in the Mediterranean area. Plant Biosyst 148(2):384–391. https://doi.org/10.1080/11263504.2013.877533

Ortega-Calvo JJ, Hernandez-Marine M, Saiz-Jimenez C (1993) Cyanobacteria and algae on historic buildings and monuments. In: Garg KL, Garg N, Mukerji KG (eds) Recent advances in biodeterioration and biodegradation. Naya Prokash, Calcutta, pp 173–203

Ortega-Calvo JJ, Ariño X, Hernandez-Marine M, Saiz-Jimenez C (1995) Factors affecting the weathering and colonization of monuments by phototrophic microorganisms. Sci Total Environ 167(1):329–341. https://doi.org/10.1016/0048-9697(95)04593-P

Otlewska A, Adamiak J, Gutarowska B (2014) Application of molecular techniques for the assessment of microorganism diversity on cultural heritage objects. Acta Biochim Pol 61(2): 217–225

Palla F (2004) Molecular biotechnologies for the characterization and the evaluation of microorganisms in degradation processes of manufactures which are of historical artistic interest. Conserv Sci Cult Herit 4:183–194

Palla F (2006) Biotechnologies for characterisation of bioaerosol in indoor environments for cultural heritage. In: D'Amico C (ed) Innovazioni tecnologiche per i Beni Culturali in Italia. Patron Editore, Bologna, pp 83–90

Palla F (2011) Characterization of microbial communities in pest colonized books by molecular biology tools. JEAR 43(2):61–67. https://doi.org/10.4081/jear.2011.61

Palla F (2020) Biotechnology and cultural heritage conservation. In: Turcanu-Carutiu D (ed) Heritage. IntechOpen, London, pp 239–252

Palla F, Anello L, Pecorella S, Russo R, Damian F (2003) Characterization of bacterial communities on stone monuments by molecular biology tools. In: Saiz-Jimenez C (ed) Molecular biology and cultural heritage. Swets & Zeitlinger BV, Lisse, pp 115–118

Palla F, Billeci N, Mancuso FP, Pellegrino L, Lorusso LC (2010) Microscopy and molecular biology techniques for the study of biocenosis diversity in semi-confined environments. Conserv Sci Cult Herit 10:185–194

Palla F, Sineo L, Manachini B (2011) Bacteria, fungi and arthropod pests collected on modern human mummies. JEAR 43(2):69–76. https://doi.org/10.4081/jear.2011.69

Palla F, Cammarata M, Trapani MR, Ghersi G, Salamone M, Sebastianelli M (2013) Novel proteases from marine organisms with potential interest in restoration procedures. In: Rogerio-Candelera MA, Lazzari M, Cano E (eds) Science and technology for the conservation of cultural heritage. Taylor & Francis Group, London, pp 279–282. https://doi.org/10.1201/b15577-65

Palla F, Figuccio B, Sebastianelli M, Vitella M (2015) Anthropomorphic wooden reliquaries an integrated approach to restoration. Eur J Sci Theol 11(2):25–32

Parulekar-Berde C, Surve RR, Salvi SP, Rawool PP, Chari PVB, Berde VB (2020) Bioremediation of cultural heritage: removal of organic substances. In: Yadav A, Rastegari A, Gupta V, Yadav N (eds) Microbial biotechnology approaches to monuments of cultural heritage. Springer, Singapore. https://doi.org/10.1007/978-981-15-3401-0_6

Pasquarella C, Balocco C, Pasquariello G, Petrone G, Saccani E, Manotti P, Ugolotti P, Ugolotti M, Palla F, Maggi O, Albertini R (2015) A multidisciplinary approach to the study of cultural heritage environments: experience at the Palatina library in Parma. Sci Total Environ 536:557–567. https://doi.org/10.1016/j.scitotenv.2015.07.105

Pasquariello G, Barresi G, Di Carlo E, Rotolo V, Palla F (2016) Non-invasive monitoring of microbial contamination on graphic collections preserved in museums, libraries and archives. In: Abstracts book of IX Congresso Nazionale AIAr, Arcavacata di Rende, 9–11 Marzo, p 3

Pavlogeorgatos G (2003) Environmental parameters in museums. Build Environ 38(12): 1457–1462. https://doi.org/10.1016/S0360-1323(03)00113-6

Perry TD, McNamara CJ, Mitchell R, Hernandez-Duque G (2003) An investigation of bacterial dissolution of Maya limestone: biodiversity and functional analysis. In: Saiz-Jimenez C (ed) Molecular biology and cultural heritage. Swets and Zeitlinger, Lisse, pp 137–140

Piervittori R, Salvadori O, Laccisaglia A (1996) Literature on lichens and biodeterioration of stone works. Lichenologist 28(5):471–483. https://doi.org/10.1017/S0024282996000606

Piñar G, Sterflinger K (2009) Microbes and building materials. In: Cornejo DN, Haro JL (eds) Building materials: properties, performance and applications. Nova, New York, pp 163–188

Piñar G, Piombino-Mascali D, Maixner F, Zink A, Sterflinger K (2013) Microbial survey of the mummies from the capuchin catacombs of Palermo, Italy: biodeterioration risk and contamination of the indoor air. FEMS Microbiol Ecol 86(2):341–356. https://doi.org/10.1111/1574-6941.12165

Piñar G, Sterflinger K, Pinzari F (2015) Unmasking the measles-like parchment discoloration: molecular and microanalytical approach. Environ Microbiol 17(2):427–443. https://doi.org/10.1111/1462-2920.12471

Polo A, Cappitelli F, Brusetti L, Principi P, Villa F, Giacomucci L, Ranalli G, Sorlini C (2010) Feasibility of removing surface deposits on stone using biological and chemical remediation methods. Microbiol Ecol 60(1):1–14. https://doi.org/10.1007/s00248-009-9633-6

Polo A, Gulotta D, Santo N, Di Benedetto C, Fascio U, Toniolo L, Villa F, Cappitelli F (2012) Importance of subaerial biofilms and airborne microflora in the deterioration of stonework: a molecular study. Biofouling 28(10):1093–1106. https://doi.org/10.1080/08927014.2012.729580

Prasad Gupta S, Sharma K (2012) The role of fungi in biodeterioration of sandstone with reference to Mahadev temple, Bastar. Chhattisgarh Recent Res Sci Technol 4(3):18–21

Prieto B, Silva B, Aira N, Álvarez L (2006) Towards a definition of a bioreceptivity index for granitic rocks. Perception of the change in appearance of the rock. Int Biodeterior Biodegradation 58(3–4):150–154. https://doi.org/10.1016/j.ibiod.2006.06.015

Pyzik A, Ciuchcinski K, Dziurzynski M, Dziewit L (2021) The bad and the good - microorganisms in cultural heritage environments - an update on biodeterioration and biotreatment approaches. Materials (Basel) 14(1):177. https://doi.org/10.3390/ma14010177

Rakotonirainy MS, Heude E, Lavédrine B (2007) Isolation and attempts of biomolecular characterization of fungal strains associated to foxing on a 19th century book. J Cult Herit 8(2):126–133. https://doi.org/10.1016/j.culher.2007.01.003

Rakotonirainy MS, Bénaud O, Vilmont L-B (2015) Contribution to the characterization of foxing stains on printed books using infrared spectroscopy and scanning electron microscopy energy dispersive spectrometry. Int Biodet Biodegr 101:1–7. https://doi.org/10.1016/j.ibiod.2015.02.031

Ranalli G, Sorlini C (2008) Bioremediation, in plant biology for cultural heritage. Biodeterioration and conservation. In: Caneva G, Nugari MP, Salvadori O (eds) The Getty conservation Institute. CA, Los Angeles, pp 340–346

Ricci S, Davidde B (2012) Some aspects of the bioerosion of stone artefact found underwater: significant case studies. Conserv Manage Archaeol Sites 14(1/4):28–34. https://doi.org/10.1179/1350503312Z.0000000003

Roldan M, Clavero E, Castel S, Hernandez-Marine M (2004) Biofilms fluorescence and image analysis in hypogean monuments research. Algologic Stud 111(1):127–143. https://doi.org/10.1127/1864-1318/2004/0111-0127

Sahab Ahmed F, Sidkey Nagwa M, Abed Nermine N (2014) Studies on indoor air quality in the repositories of the national library and archives of Egypt. Int J Sci Res 3(11):2122–2128

Saiz-Jiménez C (1993) Deposition of airborne organic pollutants on historic buildings. Atmos Environ 27(1):77–85. https://doi.org/10.1016/0957-1272(93)90047-A

Saiz-Jimenez C (1999) Biogeochemistry of weathering processes in monuments. Geomicrobiol J 16(1):27–37. https://doi.org/10.1080/014904599270721

Salvadori O (2000) Characterisation of endolithic communities of stone monuments and natural outcrops. In: Ciferri O, Tiano P, Mastromei G (eds) Of microbes and art: the role of microbial communities in the degradation and protection of cultural heritage. Kluwer Academic/Plenum Publishers, New York, pp 21–38

Salvadori O, Charola AE (2011) Methods to prevent biocolonization and recolonization: an overview of current research for architectural and archaeological heritage. In: Charola AE, McNamara C, Koestler R (eds) Biocolonization of stone: control and preventive methods proceedings from the MCI workshop series. Smithsonian Institution Scholarly Press, Washington DC, pp 37–50

Salvadori O, Municchia AC (2016) The role of fungi and lichens in the biodeterioration of stone monuments. Open Conf Proc J 7(1:M4):39–54. https://doi.org/10.2174/2210289201607020039

Sazanova KV, Shchiparev SM, Vlasov DY (2014) Formation of organic acids by fungi isolated from the surface of stone monuments. Mikrobiologiia 83(5):516–522. https://doi.org/10.1134/S002626171405021X

Sharma BRN, Chaturvedi K, Samadhia NK, Tailor PN (1985) Biological growth removal and comparative effectiveness of fungicides from Central India temples for a decade in-situ. In: Proceedings 5th international congress on deterioration and conservation of stone, Lausanne, vol 2. Presses polytechniques romandes, Switzerland, pp 675–683

Sharma K, Agrawal M, Jaiswal AK (2011) Biodeterioration of ancient monuments: problems and prospects. Cur Bot 2(1):23–24

Sobiechowska-Sasim M, Stoń-Egiert J, Kosakowska A (2014) Quantitative analysis of extracted phycobilin pigments in cyanobacteria—an assessment of spectrophotometric and spectrofluorometric methods. J Appl Phycol 26:2065–2074. https://doi.org/10.1007/s10811-014-0244-3

Souza-Egipsy V, Wierzchos J, Sancho C, Belmonte A, Ascaso C (2004) Role of biological soil crust cover in bioweathering and protection of sandstone in semi-arid landscape (Torrollones de Gabarda, Huesca, Spain). Earth Surf Process Land 29(13):1651–1661. https://doi.org/10.1002/esp.1118

Sterflinger K (2000) Fungi as geologic agents. Geomicrobiol J 17:97–124. https://doi.org/10.1080/01490450050023791

Sterflinger K (2010) Fungi: their role in deterioration of cultural heritage. Fungal Biol Rev 24(1): 47–55. https://doi.org/10.1016/j.fbr.2010.03.003

Sterflinger K, Piñar G (2013) Microbial deterioration of cultural heritage and works of art – tilting at windmills? Appl Microbiol Biotechnol 97(22):9637–9646. https://doi.org/10.1007/s00253-013-5283-1

Sterflinger K, Pinzari F (2012) The revenge of time: fungal deterioration of cultural heritage with particular reference to books, paper and parchment. Environ Microbiol 14(3):559–566. https://doi.org/10.1111/j.1462-2920.2011.02584.x

Strzelczyk AB (2004) Observations on aesthetic and structural changes induced in polish historic objects by microorganisms. Int Biodeterior Biodegradation 53(3):151–156. https://doi.org/10.1016/S0964-8305(03)00088-X

Strzelczyk AB, Bannach L, Kurowska A (1997) Biodeterioration of archeological leather. Int Biodeterior Biodegradation 39(4):301–309. https://doi.org/10.1016/S0964-8305(97)00026-7

Suihko M, Alakomi H, Gorbushina A, Fortune I, Marquardt J, Saarela M (2007) Characterization of aerobic bacterial and fungal microbiota on surfaces of historic Scottish monuments. Syst Appl Microbiol 30(6):494–508. https://doi.org/10.1016/j.syapm.2007.05.001

Swings J, Descheemaeker P (1995) The application of fatty-acid methyl-ester analysis (fame) for the identification of heterotrophic bacteria present in decaying lede-stone of the St Bavo

Cathedral in Ghent. Sci Tot Environ 167:241–247. https://doi.org/10.1016/j.syapm.2007.05.001

Szostak-Kotowa J (2004) Biodeterioration of textiles Int Biodeter Biodegr 53(3):165–170. https://doi.org/10.1016/S0964-8305(03)00090-8

Tayler S, May E (1995) A comparison of methods for the measurement of microbial activity on stone. Stud Conser 40:163–170. https://doi.org/10.2307/1506474

Tiano P (1986) Problemi biologici nella conservazione del materiale lapido. La Prefabbricazione 22:261–272

Tiano P, Tommaselli L (1989) Un caso di biodeterioramento del marmo. Ark Dent 6:211–217

Tomaselli L, Lamenti G, Bosco M, Tiano P (2000) Biodiversity of photosynthetic micro-organisms dwelling on stone monuments. Int Biodeterior Biodegradation 46(3):251–258. https://doi.org/10.1016/S0964-8305(00)00078-0

Torre MA, Gomez-Alarcon G, Palacios JM (1993) 'In vitro' biofilm formation by *Penicillium frequentans* strains on sandstone, granite, and limestone. Appl Microbiol Biotechnol 40(2):408–415. https://doi.org/10.1007/BF00170402

Troiano F, Vicini S, Gioventù E, Lorenzi PF, Improta CM, Cappitelli F (2014) A methodology to select bacteria able to remove synthetic polymers. Polym Degrad Stabil 107:321–327. https://doi.org/10.1016/j.polymdegradstab.2013.12.029

Urzì C, De Leo F (2001) Sampling with adhesive tape strips: an easy and rapid method to monitor microbial colonization on monument surfaces. J Microbiol Methods 44(1):1–11. https://doi.org/10.1016/S0167-7012(00)00227-X

Urzì C, Krumbein WE (1994) Microbiological impacts on the cultural heritage. In: Krumbein WE, Brimblecombe P, Cosgrove DE, Staniforth S (eds) Durability and change: the science, responsibility, and cost of sustaining cultural heritage. Willey, Chichester/West Sussex, pp 107–135. https://doi.org/10.13140/2.1.2289.1528

Valentin N (2003) Microbial contamination in museum collections: organic materials. In: Saiz-Jimenez C (ed) Molecular biology and cultural heritage. Swets & Zeitlinger, Lisse

Vaughan DJ, Pattrick RAD, Wogelius RA (2002) Mineral, metals and molecules: ore and environmental mineralogy in the new millennium. Mineral Mag 66(5):653–676. https://doi.org/10.1180/0026461026650054

Villa F, Pitts B, Lauchnor E, Cappitelli F, Stewart P (2015) Development of a laboratory model of a phototroph-heterotroph mixed-species biofilm at the stone/air interface. Front Microbiol 6(1251):1–14

Vukojević J, Grbić ML (2010) Moulds on paintings in Serbian fine art museums. Afr J Microbiol Res 4(13):1453–1456. https://doi.org/10.2298/ZMSPN0519175V

Wagner E, Schwartz W (1965) Geomikrobiologische Untersuchungen. IV Untersuchungen über die mikrobielle Verwitterung von Kalkstein im Karst. Z Allg Mikrobiol 5(1):52–76. https://doi.org/10.1002/jobm.19650050108

Warscheid T, Braams J (2000) Biodeterioration of stone: a review. Int Biodeterior Biodegradation 46(4):343–368. https://doi.org/10.1016/S0964-8305(00)00109-8

Woese CR (2000) Interpreting the universal phylogenetic tree. Proc Natl Acad Sci U S A 97(15):8392–8396. https://doi.org/10.1073/pnas.97.15.8392

Wollenzien U, de Hoog GS, Krumbein WE, Urzí C (1995) On the isolation of microcolonial fungi occurring on and in marble and other calcareous rocks. Sci Total Environ 167(1–3):287–294. https://doi.org/10.1016/0048-9697(95)04589-S

Zanardini E, Abbruscato P, Ghedini N, Realini M, Sorlini C (2000) Influence of atmospheric pollutants on the biodeterioration of stone. Int Biodeterior Biodegradation 45(1–2):35–42. https://doi.org/10.1016/S0964-8305(00)00043-3

Bioaerosol

2

Paola De Nuntiis and Franco Palla

Abstract

Cultural heritage constitutive materials can provide excellent substrates for microbial colonisation, highly influenced by thermo-hygrometric parameters. In cultural heritage-related environments, a detrimental microbial load may be present both on artworks surface and in the aerosol. Confined environments (museums, archives, deposits, caves, hypogea) are characterised by peculiar structures and different thermo-hygrometric conditions, influencing the development of a wide range of microbial species, able to induce artefact biodeterioration and to release biological particles in the aerosol (spores, cellular debrides, toxins, allergens) potentially dangerous for the human health (visitors/users). In order to identify the real composition of the biological consortia, highlighting also the symbiotic relationships between microorganisms (cyanobacteria, bacteria, fungi) and macro-organisms (plants, bryophyte, insects), an interdisciplinary approach is needed.

The results from in vitro culture, microscopy and molecular biology analysis are essential for a complete understanding of both microbial colonisation of the cultural objects and the potential relationship with illness to human. Concerning the bioaerosol, of crucial importance are the time and techniques for sampling.

P. D. Nuntiis (✉)
Italian National Research Council – Institute of Atmospheric Sciences and Climate (CNR-ISAC), Bologna, Italy
e-mail: p.denuntiis@isac.cnr.it

F. Palla
Department of Biological, Chemical, Pharmaceutical Sciences and Technologies, Laboratory of Biology and Biotechnologies for Cultural Heritage, University of Palermo, Palermo, Italy
e-mail: franco.palla@unipa.it

F. Palla, G. Barresi (eds.), *Biotechnology and Conservation of Cultural Heritage*,
https://doi.org/10.1007/978-3-030-97585-2_2

Keywords

Biodeterioration · Microbial colonisation · Preventive conservation

2.1 Indoor Environments (Libraries, Museums, Storerooms, Hypogea, Churches)

Several aspects of indoor environments need to be defined. Their main function is to preserve the objects, made of different materials that make up a country's historical and artistic cultural heritage, in addition to serving educational purposes and representing cultural and social identity. Indoor environments may either be constructed in situ to preserve artworks in their place of origin (hypogea) or custom-built using innovative techniques and with air-conditioned premises.

Alternatively, they may be realised in existing monumental buildings which constitute artworks themselves, but also house works of art. The type of building structure will determine the risk and potentially the type of deterioration and problems associated with managing the indoor environment and collections.

In turn, the location of indoor environments, in urban or rural centres, industrialised or green areas and coastal or mountain regions (Thomson 1986; Camuffo 2019) will determine their different external and internal microclimatic conditions, the aerosol composition and its biological impact. As indoor air is linked to the atmosphere surrounding the building (Brimblecombe 1990), also a wide number of visitors will have a negative effect on the indoor environment. Moreover, the shape of indoor spaces may have a further influence, as they can range from large, very high premises to small interconnecting areas arranged either on several floors or on a single level, attics or basements, with or without windows, varying exposure, adjoined or separate, with showcases, display cabinets or clima boxes representing micro-environments enclosed within the main exhibition macro-environments (Michalski 1994; De Guichen and Kabaoglu 1985; De Guichen 1980; Cassar 1995; IBC 2007; Lazaridis et al. 2015).

Indoor microclimatic conditions deeply depend on whether active heating/air-conditioning indoor and lighting systems are present or not in a given environment. Most indoor environments, especially those constructed in loco (hypogeum), or in historical buildings or premises mainly destined for other uses (churches), lack active microclimate control systems. Even when conditioning systems are present, they are often adjusted to the needs of human comfort on the part of museum staff or church worshippers rather than those of artefact conservation. In addition, unless these systems are regularly serviced, they can be a further source of potential chemical and, above all, biological pollutants (Sedlbauer 2002; Varas-Muriel et al. 2014).

The study and control of the microclimate (Cavallini et al. 1991; IBC 2007; Bernardi 2009; Camuffo 2019) is essential to curb "biological risk", as microclimatic conditions have a major impact on spore germination. The main indoor microclimatic parameter is relative humidity. Very high relative humidity levels in the winter months, or in conditions of moist warm wind, may result in condensation followed

Fig. 2.1 Hypogeum archaeological site, in addition to the glass ceiling (*dark area at the top*), allowing natural light to penetrate, two older-generation lighting systems are visible: fluorescent tubes lacking UV guards and halogen spotlights generating considerable heat and relative air movement. This lighting system combined with high relative humidity levels increases the risk of biodegradation. The site is currently under monitoring and closed to the public

by microbial colonisation on the internal or external surfaces of a building and even worse on the frescoes of the masonry walls (Camuffo 2007; Ranalli et al. 2018).

Under these conditions, churches are more at risk of biodeterioration than hypogea due to the different materials they contain ranging from stone architectural and decorative features to organic materials like wood used for floorings, ceilings and furnishings and objects linked to worship (Nugari 2003). The thermo-hygrometric excursions during daily liturgical celebrations typical of the mountain churches, have not to be overlooked (Camuffo et al. 1999). With the possible exception of storerooms, another problematic parameter in all these indoor environments is natural or artificial light. Light heats the illuminated surface generating hygrometric and mechanical stress and warms the circulating air mass triggering convection that not only enhances the inertial deposition of suspended particulate matter but also pigment discoloration and biodegradation (Fig. 2.1). Light is needed to display objects, but existing lighting systems are often outdated and obsolete. Nonetheless, a variety of lighting systems are currently available that ensure optimum object preservation, saving both energy and money, including through the use of sensors that trigger light only when visitors enter an environment (IBC 2007).

Biological contamination can also occur in objects newly acquired from uncontrolled indoor environments (donations from private collections or excavated materials in contact with the soil), but is also common after natural or man-made disastrous events like flooding. Visitors also act as carriers of outdoor spores transported inside the building through clothes, skin and hair, they contribute to

Table 2.1 Average contribution ascribed to individual museum visitors

Contribution	Cause	Quantity
RH—water vapour	Breathing, sweating	Approx. 40–100 g/h
T—heat	Movement	Approx. 100 W/h
CO_2—carbon dioxide	Breathing	Approx. 20 l/h
Dust		
Fibres	Clothing	Approx. 0.2 g/m^3 h
Microorganisms	Sweating, talking, sneezing	
Organic fragments	Skin, hair	

Cited by Mandrioli (2015)

increasing CO_2 linked to breathing, temperature, water vapour and microorganisms through sweating, talking and sneezing (Table 2.1). Indoor exhibition/conservation environments such as museums and churches often contain different materials and multi-material artworks, which give rise to a number of conservation problems linked to the intrinsic features of the materials (Hueck 2001) and their microclimatic requirements for conservation and management. Related to this aspect are storerooms located in attics or basements, which are often poorly maintained, full of dust and insects, and have suboptimal ventilation and unstable microclimatic conditions due to their location and frequent lack of air-conditioning systems. These environments are often neglected in terms of conservation measures as they are deemed less important, despite housing large quantities of precious artefacts made of different materials for long periods of time. In addition, storerooms often house restored works of art that emit further pollutant gases generated by restoration treatments (Cappitelli et al. 2004).

Artworks dust protection has been recently achieved using the Compactus shelves, but these microenvironments can cause fungal colonisation, especially in historical libraries on leather or fabric books binding (Micheluz et al. 2015). A regular and adequate cleaning, is a highly recommended preventive conservation strategy in every environment (Getty Conservation Institute 1994); in fact, humidity associated with mild-moderate temperatures and accumulated dust may create a favourable habitat for a biodeterioration of many materials (Sterflinger and Pinzari 2012). In the already contaminated material, moisture is well preserved in the biofilm, so even conditions of RH $>$ 50% are sufficient for the survival of some microorganisms (Camuffo 2019).

Finally, a complete analysis should not neglect the assessment of building vulnerability, considering the physical and managerial criticalities related to poor maintenance of roofs, windows and masonry (Bonazza et al. 2021; Cacciotti et al. 2021, Sesana et al. 2021), especially caused by the extreme hydrometeorological phenomena, due to climate change and closely connected to the microclimate and indoor microbial exposure. The use of water-sensitive building materials in areas with hot and humid climates can also lead to indoor mould growth.

Environmental monitoring continues to be the main tool not only for recording but above all for actively keeping under control the causes of material degradation. Based on a series of data recorded over a sufficiently long period of time, continuous monitoring will permit to correctly analyse indoor and outdoor environmental measurements based on specific targets. The importance of on-site monitoring is also reiterated in the Sendai Framework for Disaster Risk Reduction 2015–2030 (Sendai Framework - United Nations 2015)[1] Priority 4 (Enhancing disaster preparedness for effective response and to "Build Back Better" in recovery, rehabilitation and reconstruction) in the sectors of risk due to climate change which also includes Cultural Heritage (Bonazza et al. 2018). The knowledge of climate change impact on cultural heritage in Italy is principality based on the identification of the main climatic parameters for degradation of indoor and outdoor artworks. A specific threat to cultural heritage concerns changes in biodegradation processes due to projections of seasonal rainfall (National Plan for Adaptation to Climate Change—PNACC[2] which makes effective the National Strategy for Adaptation to Climate Change—SNACC[3]).

2.2 Airborne Particles: Organic and Vegetable Dust and Biohazards

Bioaerosol is the scientific term used to define a suspension of aerosols or particulate matter of microbial, plant or animal origin and may consist of pathogenic or non-pathogenic, live or dead bacteria and fungi, algae, viruses, pollen, plant fibres, high molecular weight allergens, bacterial endotoxins, mycotoxins, peptidoglycans or glucans (Douwes et al. 2003) passively carried by air (Cox and Wathes 1995). For this reason, the term bioaerosol does not include insects (Mandrioli and Ariatti 2001).

Airborne biological material is composed of particles generated from natural sources by active or passive mechanisms and resuspended in the atmosphere as individual organisms or often aggregated with each other or with non-biological solid or liquid particles in suspension (Lighthart and Stetzenbach 1994). The biological aerosol includes many types of airborne particles varying widely in morphology and size that can be seen with a magnifying glass or microscope. Their aerodynamic diameter may range from molecular size to large and giant particle size (Jaenicke 2005; Hinds 1999; Pöschl 2005), e.g. viruses (1 nm–1 μm), bacteria (0.1–1 μm), fungal spores (0.5–50 μm), lichen propagules (10 μm–1 mm), bryophyte spores (1–100 μm), algal cells (1 μm–1 mm) and pollen grains (10–100 μm) (Després et al. 2012; Fuzzi et al. 2015). Airborne particles are

[1] https://www.undrr.org/implementing-sendai-framework/what-sendai-framework

[2] https://www.minambiente.it/sites/default/files/archivio/allegati/clima/pnacc.pdf

[3] https://www.minambiente.it/notizie/strategia-nazionale-di-adattamento-ai-cambiamenti-climatici-0

associated with other biological materials such as protozoan cysts found in aggregates or incorporated in solid or liquid particles, pteridophyte spores, plant fragments and products from metabolic activities. Another term commonly used to describe airborne particles derived from biological organisms is *primary biological aerosol particles* (PBAPs) that differentiates biogenic particles from secondary organic aerosols formed by further physical processes and chemical reactions in the atmosphere such as photo-oxidation (Després et al. 2012). When microorganisms are incorporated, for instance, into fog droplets for many hours, the environmental conditions are favourable for rapid growth, giving rise to *secondary biological aerosol particles*, commonly called SBAPs (Fuzzi et al. 1997, 2015; Després et al. 2012).

Aerobiology is a relatively new scientific discipline specifically dealing with airborne particles, how they behave in the air once generated, how the environment influences their dispersion and deposition and the impact these particles have on other organisms or materials such as artworks. The presence of bioaerosol in the atmosphere is strictly correlated to an active source able to produce material through physiological processes generating microorganisms or physical processes resulting in disaggregation and fragmentation of organisms. Sources of bioaerosol emission can be natural, for example, bacteria found in the air often belong to groups commonly present in the ground, and in fresh and sea water, while bacteria, algae and fungi are released into the air by a bubble-bursting mechanism influenced by the wind. Forests and vegetation are sources of pollen, spores and fragments, while anthropogenic sources include farming and agricultural processing and industrial activities.

The particles produced are emitted into the atmosphere by both physiological (e.g. the catapult expulsion mechanism for the dispersal of *Parietaria* (*officinalis, judaica, lusitanica, creatica*) allergens and physical processes (e.g. fungal spores released by the action of wind and rain on vegetation) (Mandrioli 1985). The process of fungal spore release mainly depends on atmospheric agents, relative humidity, air temperature, dew point temperature and wind turbulence (Jones and Harrison 2004). Peak fungal spore concentration in temperate and Mediterranean climates coincides with summer and autumn when relative humidity is higher but is strongly influenced by the frequency of precipitations that attenuate relative humidity values. Seasonal variations in pollen production have a major impact on the ratio between indoor and outdoor spore concentrations. In spring and summer, the peak concentration of most fungal spores detected in indoor environments is similar to that found outside, whereas in winter the indoor concentration is higher (Mandrioli et al. 1998; Sabbioni et al. 2008), thereby confirming the trapping effect of buildings. Post-injection conditions in the atmosphere are due to the survival of microorganisms and are controlled by physical and chemical parameters. The dynamic nature of the atmosphere makes difficult to assess the individual importance of each parameter. The major limiting factors are temperature, relative humidity, ultraviolet radiation, oxygen, carbon monoxide, nitrogen dioxide and formaldehyde (Mandrioli 1998). Although bioaerosol is ubiquitous its concentration in the atmosphere varies with season and location, and it has been estimated to constitute up to 25% of total aerosol mass and is sometimes numerically close to 50% of all aerosol particles on a global basis (Jones and Harrison 2004; Jaenicke 2005; Walser et al. 2015).

Recent studies implementing molecular techniques demonstrated that the fungal spectrum suspended in the air is much richer than previously known (Fröhlich-Nowoisky et al. 2009; Després et al. 2012; Pashley et al. 2012).

The transport or dispersal of bioaerosol in the atmosphere is a physical process based on kinetic energy exchanged when gas particles in the atmosphere clash with motionless air-dispersed particles. Particles do not behave consistently so that each stage is random in both duration and direction. Due to unfavourable environmental conditions such as dehydration and UV radiation (Griffith and De Cosemo 1994), the atmosphere contains not only vegetative forms but many forms of resistance like bacterial and fungal spores. Small particles, ranging in size from 1.0 to 5.0 μm, remain suspended in the air for a longer time, whereas larger particles tend to settle more quickly on surfaces due to their larger mass. Bioaerosol can be transported in the atmosphere for long distances (Gregory 1973; Schlesinger et al. 2006) and a longer time due to its vicinity to particle sources and resuspension of deposited particles (Tampieri et al. 1977; Mandrioli et al. 1980, 1984; Rantio-Lehtimäiki 1994). Kellogg and Griffin (2006) identified the global transport of desert dust as the main mechanism responsible for the transport of aerosol microbiota: pollens, fungi and bacteria. The average residence time of biological particles in the atmosphere can range from less than a day to a few weeks, depending on their size and aerodynamic properties (De Nuntiis et al. 2003; Després et al. 2012).

Particle deposition is the aspect of most interest to cultural heritage and, in particular, the biodeterioration of artworks. However, bioaerosol in the atmosphere is only one of the potential risks arising when deposited material encounters favourable environmental conditions for the colonisation of artefacts. Particles are usually removed from the air by sedimentation and deposition on all surfaces, not only horizontal planes. Deposition occurs by gravitational settling, molecular diffusion and impact. A highly effective but discontinuous means of bioaerosol removal in outdoor environments is rainout and washout, which happens during precipitations when damp deposition captures the particulate in precipitations and deposits it on the ground. Precipitation is the most efficient removal mechanism for particles 0.1–10 μm in diameter. Computation of bioaerosol deposition velocity is a complex problem as particles are irregular in shape and their structural features hamper calculation of particle density, e.g. despite its size, the two air bladders of *Pinus* pollen grains make them particularly light, thereby increasing dispersal distance (Schwendemann et al. 2007). In addition, particles vary in relation to atmospheric humidity changes: small particles are dispersed among air molecules increasing in velocity, whereas large particles shift the surrounding air creating vortices and falling more slowly. Particle deposition is slowed down if the descending particle trajectory is close to a vertical surface, whereas the velocity changes when particles are clustered together. Particles are also affected by thermophoresis and diffusiophoresis, temperature and concentration gradients and electrostatic forces that not only induce particle accumulation with blackening of the surfaces involved but also biodeterioration when conditions are favourable. Particles settled by dry or wet deposition can be involved in resuspension mechanisms and hence return once again into the atmosphere. Once deposited, biological particles can interact with the substrate, be it the nasal mucosa, a leaf surface or a fifteenth-century fresco, giving rise to an allergic reaction in sensitive patients, a plant disease following fungal or bacterial colonisation or mechanical and aesthetic deterioration of a painted surface. Fungal attack of a fresco surface, due to its porosity, can lead to hypha penetration of the painted layer resulting in flaking and detachment of the fresco surface, coloured stains obscuring the painting and the production of acid metabolites or enzymes able to transform complex molecules into simple water-soluble molecules. Over time, this process will weaken the painted layer damaging the material and value of the artwork. Studies on biodeterioration are a constantly evolving field for scientific and technological research. They currently focus not only on bioaerosol sampling and identification methods but also on transport and deposition mechanisms and above all on the ecology of the species involved.

2.3 Impact of Colonised Aerosol on Artwork Surfaces and Potential Enemies of Human Health

Studies on the biodeterioration of cultural heritage are not confined to the microbiology of biodeteriogens and material degradation processes; they require a multidisciplinary approach to understand the chemical and biological relations between the air and the materials it surrounds. Research in this sector not only serves to identify potential risk factors for artwork preservation, devise specific preventive conservation protocols or to identify the most suitable treatments, but it also serves to safeguard the health of operators (conservationists, restorers) and visitors from the risk of exposure through inhalation and contact with contaminated surfaces or objects. The risk for human health stems not only from harmful microbial species or the products of their metabolic activity, e.g. allergens, present on artefacts or in the air, but also from the hazardous residues of biocide treatments used in the cultural heritage sector. Another aspect, neglected in the past, is that some components of the bioaerosol as well as some environmental factors may have synergistic effects; for instance, you may think of the hypothesis in the health sector that hypersensitivity pneumonitis can develop after infection with the influenza virus (Gudmundsson et al. 1999). Indeed research on SARS-CoV-2 virus and COVID-19 disease has been strongly oriented, as never before, to study environmental and biological synergies, having impacts that will also affect the field of cultural heritage protection. Bioaerosol research started thanks to the interest of the health sector (allergology) and agriculture (phytopathology). The cultural heritage sector is just one of many areas in which bioaerosol can cause damage to persons and/or objects, with major economic and other related consequences. The biodegradation of cultural heritage, therefore, has a primary cultural, scientific and economic impact. As previously mentioned, bioaerosol on artworks is only harmful for preservation in concomitance with other factors: microclimatic conditions, the nature of the object, its state of preservation and chemical and physical degradation processes already in place.

Biological degradation is seldom caused by a single microorganism but is produced by complex communities, real ecosystems that develop on the artefact. Under favourable microclimatic conditions (RH, T and light), the bioaerosol deposited on the surface can grow and reproduce itself using the substrate as a nutrient (heterotrophic) or support (autotrophic), causing damage to the material component (Hueck 2001) of the cultural heritage whether it consists of traditional materials and/or modern materials like polymers. The substrate may be a statue, painting, old parchment, cave painting, glass window, liturgical vestment or fresco, made of a single organic or inorganic material or several different materials combined. The inorganic substrate can provide microorganisms with an exclusive supply of mineral salts and a limited amount of water depending on the material's porosity. The microorganisms colonising these artworks are therefore photo- or chemoautotrophic, i.e. able to self-synthesise the molecules required for their development by photosynthetic reaction (photosynthesis) or chemical reactions (chemosynthesis). Heterotrophic species can only penetrate the same artwork after autotrophic organisms, whereas organic materials are colonised by heterotrophic bacteria able to utilise the nutrients available in the material itself. The organisms causing damage to works of art are called biodeteriogens, but do not correspond to all the bioaerosol deposits generally found on them. Some of these particles may be viable but not culturable as they form colonies on solid media under certain growth conditions (time, temperature and nutrients). Many bioaerosol particles cannot be cultured on conventional media, but their existence can be proved using other methods (Năşcuţiu 2010; Blais-Lecours et al. 2015). An indoor environment (museums, galleries, archives, churches and hypogea) can be particularly suitable for microbial growth as it protects the microorganisms themselves from extreme variations in outside temperature and UV rays that can damage the bioaerosol. Desiccation, radiation, oxygen, ozone and its reaction products together with various pollutants can operate cumulatively affecting the viability of microorganisms (Griffith and De Cosemo 1994). For this reason, biological spores survive better in air than vegetative cells, as the humidity in the air is an important potential source of microorganism stress. Fungi are among the most harmful organisms associated with the biodeterioration of organic and inorganic materials (Sterflinger 2010). Many organisms excrete waste metabolic products, including pigmented or acid compounds that may disfigure materials, altering their colours or causing mechanical damage. The presence of heavy metals in some pigments used in paintings on the other hand can increase the resistance to microbial deterioration (Ciferri 1999).

Aesthetic biodeterioration should not be underestimated as it severely alters the perception of beauty and the legibility of artworks although generally it is less aggressive towards the materials. For example, the fungi present on different types of materials can determine the formation of visible films, spots, exfoliation, disruption and pitting. Chemical processes lead to the transformation, alteration and decomposition of the substrate and are much more common than in the past. In addition, the pores and fractures caused by chemical and mechanical interactions can host further biodeteriogens (Urzì et al. 2000). A chemical action may be attributed both to assimilation processes (when the organisms use the material as nourishment by means of extracellular enzyme activity or ion exchange) and to the excretion of metabolic intermediates or substances having an inhibitory or waste function (such as acid, alkaline and complexing substances and pigments). In physical biodeterioration, the organism breaks or simply deforms the material with growth or movement. Although there are many studies in the literature on microbial contamination relating to works of art preserved in different structures such as museums, crypts, churches, hypogea, libraries and archives (Valentin 2003; Nugari et al. 2003; Gaüzère et al. 2014; Tarsitani et al. 2014; Kavklera et al. 2015; Ruga et al. 2015; Ranalli et al. 2018), there is still much to do on the definition of danger thresholds for biodeterioration processes.

2.4 Revealing and Identifying Microbial Particles and Products

No automatic instruments are currently available for the direct measurement of viable and non-viable microorganisms in the air or on surfaces (Caneva et al. 2020). Nor is there a universal bioaerosol sampler: available devices must provide a representative sample trying to minimise stress (e.g. dehydration) and damage to the biological activity of microorganisms. The main aim of sampling is to identify the biological particles (qualitative sampling) presents in the sampled location and measure the variation of their concentration in the atmosphere (quantitative sampling).

Before embarking on sampling, it is important to establish what is being sought and where, which is the best sampling point in relation to the environmental characteristics and the presumed degree of contamination so as to set appropriate sampling times. Alongside biological sampling, it is particularly important to undertake parallel sampling of the main physical and chemical environmental parameters. Last but not least, the most suitable analysis techniques must be chosen to identify and quantify the bioaerosol isolated, especially in the case of viable material. It should be emphasised again that the exact bioaerosol fraction cannot be known using only culture-dependent methods since microbes may be viable but non-cultivable, underestimating both microbial diversity and concentration (Blais-Lecours et al. 2015). Nevertheless, the combination of culture-dependent and molecular analysis should be promoted to increase the observed microbial diversity (Palla 2015; Saiz-Jimenez and Gonzalez 2007) and should be adopted also in the field of cultural heritage diagnostics.

2.4.1 Sampling by Passive or Active Methods

The simplest technique, and hence the most commonly used by non-experts, is gravitational deposition, exposing a horizontal surface on which particles settle by gravity and remain trapped by an adhesive placed on the sampling surface or directly on a semisolid culture. Sampling efficacy will depend on air conditions, wind direction and speed, as well as particle concentration and dimensions. The passive

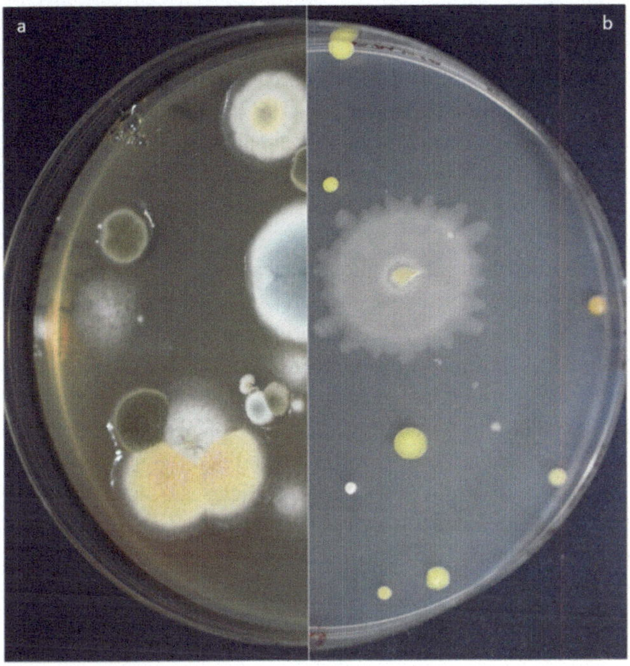

Fig. 2.2 Passive sampling carried out in the same indoor environment using (**a**) Sabouraud medium (fungal colonies) and (**b**) Nutrient agar medium (bacterial colonies)

sampling allows a qualitative investigation, as the volume of sampled air and the efficiency of capture are not known. It is only suitable for undisturbed indoor environments and is also used for scientific tests in Italian heritage sites, such as the Sistine Chapel in Rome, the Palatina Library in Parma and in the crypt of St. Peter in Perugia (Montacutelli et al. 2000; Pasquarella et al. 2015; Ruga et al. 2015). Petri dishes measuring 90 cm in diameter, containing semisolid culture media (Sabouraud or Nutrient agar), are normally left open to the air for 1 h at 1 m from the floor and 1 m away from walls and then incubated at 30 °C for 16–72 h reaching bacterial or fungal colonies (Fig. 2.2). Culture plate sampling results first need to undergo culture analysis and are then expressed in colony-forming units per surface area (CFU/dm^2). To estimate microbial air contamination, an index of microbial air (IMA) is used, based on the count of microbial fallout on Petri dishes, expressed in CFU/dm^2/h or CFU/h (Pasquarella et al. 2000).

A sterile glass slide treated with adhesive can be used to sample non-viable bioaerosol with subsequent direct observation of the particulate under an optical microscope, the results are expressed in this case in particles per surface area (particles/cm^2). In addition to air sampling the affected surfaces of cultural objects can be sampled in a non-destructive and non-invasive way, using strips of adhesive tape (Fungi-Tape™) or sterile cotton swabs or nitrocellulose membrane filters (Sartorius AG, Göttingen, Germany), followed by observation under optical microscope or scanning electron microscope (SEM) after adequate sample preparation. The nitrocellulose membrane filters consisted of a 47 mm square disc pressed onto the sample surface for 30 s and then transferred to Petri dishes containing agar for bacteria or fungi isolation (Pitzurra et al. 1997; Pasquarella et al. 2015, Micheluz et al. 2015). To avoid material loss in very small sampling points it is possible to perform a slight aspiration, using a vacuum pump for a short time, in order to guarantee the sufficient quantity of material for subsequent analyses (Paiva de Carvalho et al. 2019).

The Andersen microbial air sampler is a cascade impactor commonly used in diagnostics in the cultural heritage sector as it samples viable bioaerosol (Andersen 1958). The device is particularly suited to indoor sampling as the aspiration tube cannot be oriented according to the wind direction. The sampler has six or three stages in which the particles are separated by size and collected on Petri dishes containing culture medium. Each stage contains plates with 400 precision-drilled holes of decreasing diameter. The slower air speed in the first stages allows the larger particles to be captured, while the smaller particles are accelerated thanks to the narrower diameter of the holes through which they must pass before being captured by the agar-treated surface. The Petri dishes can then be removed and incubated to allow the captured microbes to grow for subsequent quantification and identification. Airflow is sampled at 28.3 l per minute, wall deposition is negligible and the particle capture rate is close to 100%. Sampling time depends on bioaerosol concentration but is generally in the order of a few minutes. The only limitation of this type of sampler is the high number of dishes generated during each sampling.

The Surface Air System (SAS Super ISO, PBI International, Milan, Italy) is a much more manageable and practical portable single-stage impactor specifically designed for indoor environments of hospitals and pharmaceutical and food companies. At the international level, it is currently considered the reference instrument for microbiological air sampling with an environmental bioaerosol capture rate of 100%. The device aspirates the air at a constant flow for periods varying from a few seconds up to an hour depending on the microbial contamination. The SAS SUPER ISO 100 (180 l of air per minute) is commonly adopted in the cultural heritage sector using 55 mm diameter contact plates, but if high fungal contamination is expected, maxi 84 mm plates or 9 cm Petri dishes can be used with a special adaptor because moulds tend to spread and consequently make counting difficult after incubation. The number of colonies counted on the surface must first be corrected for the statistical possibility of multiple particles passing through the same hole, and then the CFU per cubic metre of air sampled can be calculated.

It must be mentioned that the choice of a specific microbiological growth medium, the incubation temperature and the cultivation time result to differences on the selection of the viable, cultivable, airborne microbial community. Media can be specially prepared in the laboratory or purchased ready-to-use to allow a comparison of results. This is a key feature in the cultural heritage sector where national technical standards provide indications, but standardisation is still a long way off. Tryptic soy agar (TSA) is the medium most commonly used for a total microbial count (incubation for 24–48 h a 32 °C), while Sabouraud dextrose agar (SDA) is

Fig. 2.3 Crypt environment. (**a**) Aerosol active sampling by portable sampler equipped with sterile disposable gelatine filter; (**b**) nutrient medium inoculated by Gelatine filter (completely water-soluble); (**c**) dissolving of gelatine filter during the contact with Nutrient Agar (the filter completely disappears in a few minutes)

used to count yeasts and moulds and usually includes an antibiotic (e.g. chloramphenicol) to avoid bacterial growth (incubation for 5 days at 25 °C).

Other SAS samplers are based on the capture of microorganisms by membrane filtration (SAS dust) or liquid filtration (SAS PCR) which allows identifying the material collected by means of a real-time PCR method. The air samplers that operate in impaction mode (AirPort MD8, Sartorius) collect airborne microorganisms by suctioning a defined air volume through a gelatine membrane filter (Fig. 2.3a) or a culture agar plate. Gelatine filters (Sartorius, AG Germany) allow the material collected to be transported to a culture medium for incubation as shown in Fig. 2.3b, c (Di Carlo et al. 2016).

Aspiration samplers are frequently used for the non-biological fraction of atmospheric particulate adopting membrane filters with pores of a few micron for non-viable bioaerosol (pollen grains and spores) where microorganisms remain trapped on the surface and can be examined under the light microscope either directly or after diaphanization, depending on the filter matrix.

Hirst impact sampler is used to measure the concentration trends of bioaerosol components like pollen, fungal spores, algae and other particles morphologically recognisable under optical microscope (Hirst 1995). The collection surface can be a microscope slide (for daily sampling) or a transparent plastic tape (for 7-day monitoring) treated with the application of a silicon layer to retain the impacted particles. Sampling efficiency is around 95% for particles with a diameter larger than 20 μm and 50% for particles with diameters 2–5 μm. The 2 × 14 mm slit allows the sampled air to move over a surface at 2 mm per hour so that the time trend of particulate concentration can be measured by subsequent light microscopy

Observation. The sampler's suction rate is 10 l per minute, equivalent to the average human breathing rate (Mandrioli et al. 1998), and for this reason is principally used in the field of health, in centres belonging to aeroallergen monitoring networks worldwide, performing the continuous monitoring of pollens and fungal spores (EAN, RIMA, REA, RNSA, NAB, etc.)[4] by a procedure already standardised in Italy (UNI 11108:2004)[5] and in Europe (CEN EN 16868:2019).[6]

The resulting data bank could be an interesting source of information on the daily concentrations of outdoor fungal spores in urban and rural environments for the cultural heritage sector. Three manufacturers (Lanzoni S.r.l., Bologna, Italy, Burkard Manufacturing Co. Ltd., Hertfordshire, England, Cavazza Sas, Bologna, Italy) currently adopt international recommendations to manufacture the commonly used outdoor samplers (VPPS 2000, volumetric spore trap and easy SPT100) and the portable version more often used indoors (VPPS 1000, indoor volumetric spore trap and easy SPT100 light).

2.4.2 Biochemical and Biomolecular Techniques

The analytical approach aimed at identifying the biological particles in the indoor air environment that can represent biodeterioration and health hazards includes a broad spectrum of methods. They are based on both conventional microbiological procedures and advanced techniques of molecular biology (Letch 2016; MacNeil et al. 1995). The identification of colonies isolated from air samples can be performed by observing the morphological features according to different manuals or identification keys. In many cases, however, it may be necessary to identify microbial consortia using specific staining methods (e.g. Lugol's staining, Gram staining), biochemical tests (e.g. enzymatic assay, metabolite profiling, ATP bioluminescence assay) and molecular analysis (Di Carlo et al. 2016; Lavin et al. 2014; Šimonovičová et al. 2015; Sanmartín et al. 2016; Castillo et al. 2016).

The application of molecular methods has allowed cultivation-independent investigations of microbial communities in diverse environments. Since not every microorganism in a microbial community can be isolated or cultivated, extraction and sequencing of total microbial DNA are useful to identify those microorganisms defined "not culturable" (Puškárová et al. 2016). Culture-independent methods (CIMs) are based on genetic identification (qualitative analysis) of bacteria and fungi as well as DGGE profiling and PCR that have been developed to study microbial communities from various environments (Šimonovičová et al. 2015;

---

[4]EAN, European Aeroallergen Network; RIMA, Rete Italiana di Monitoraggio in Aerobiologia; REA, Red Española de Aerobiología; RNSA, Réseau National de Surveillance Aérobiologique (France); National Allergy Bureau (USA).

[5]UNI 11108:2004—Method for sampling and counting airborne pollen grains and fungal spores (UNI, Italian standardisation body).

[6]CEN EN 16868:2019—Ambient air—Sampling and analysis of airborne pollen grains and fungal spores for networks related to allergy—Volumetric Hirst method.

Letch 2016). The strategy of PCR-mediated amplification of targeted sequences, followed by sequencing and comparative data analysis, has been used successfully on samples from air (Palla et al. 2014; Pasquarella et al. 2015). Similarly, non-PCR-based molecular techniques, such as microarray and *fluorescence* in situ *hybridisation*, have also been adopted (Su et al. 2012).

Molecular fingerprinting techniques (bacteria and fungi quantitative PCR, capillary electrophoresis single-strand conformation polymorphism fingerprinting) have been applied by many authors to analyse airborne bacteria and fungi in enclosed spaces, also in relation to bioaerosol in outdoor air, and the influence of microclimate parameters and total dust content on microbial contamination (Gaüzère et al. 2013; Skóra et al. 2015). Studies utilising culture-independent analyses of microbial communities in indoor environments give a complete overview, also based on their level of detail in documenting built environment data (Ramos and Stephens 2014).

Recent studies in DNA sequencing techniques have been carried out focusing on airborne and dust-borne microorganisms in selected museum, archive and library environments. An analytical approach using molecular fingerprinting has been applied to monitoring and characterising the airborne microbial diversity in the Louvre Museum over a long period of time (Gaüzère et al. 2014).

Microbiological contamination has been analysed in several Polish museums, libraries and archives by Skóra et al. (2015). The resulting nucleotide sequences of the identified microorganisms were analysed and compared to the sequences published in the National Center for Biotechnology Information (NCBI) database, using the BLASTN programme, confirming genetically the identified bacteria and yeasts that were previously macroscopically and microscopically characterised using Gram staining and catalase and oxidase tests. Combining cultivation-independent and cultivation-dependent studies, the fungal diversity in indoor environments was performed in order to shed light on the components of microbial consortia (Micheluz et al. 2015; Ortega-Morales et al. 2016). A widespread fungal infection was revealed in *compactus shelves* of Venetian library by Micheluz et al. (2015); particularly, xerophilic fungi were identified using a polyphasic approach based on morpho-physiological features and molecular studies. Molecular identification was performed by amplification and sequencing of internal transcribed spacers (ITS) of β-tubulin and actin genes. Moreover, airborne fungi possess great enzymatic potential to degrade materials, so their hydrolytic activity in attacking, proliferating and degrading these important artistic-historical items can be successfully detected using enzymatic assays and can be considered valuable data in completing the typical identification list of isolated strains. Recently, the biodegradative action of fungal microflora from mummified remains and fungal airborne communities was investigated using hydrolytic assays (Šimonovičová et al. 2015). Borrego et al. (2012) have determined indoor air quality in Argentine archives and the biodeterioration of documentary heritage using an analytical approach based on the qualitative determination of enzymatic fungal activity and acid production by fungi.

References

Andersen AA (1958) New sampler for the collection, sizing, and enumeration of viable airborne particles. J Bact 76(5):471–484

Bernardi A (2009) Microclimate inside cultural heritage buildings. Il Prato, Padova, p 171. ISBN: 978-88-6336-032-5

Blais-Lecours P, Perrott P, Duchaine C (2015) Non-culturable bioaerosols in indoor settings: impact on health and molecular approaches for detection. Atmos Environ 110:45–53. https://doi.org/10.1016/j.atmosenv.2015.03.039

Bonazza A, Maxwell I, Drdácký M, Vintzileou E, Hanus C, Ciantelli C, De Nuntiis P et al (2018) Safeguarding cultural heritage from natural and man-made disasters. A comparative analysis of risk management in the EU. ISBN 978-92-79-73946-0. https://doi.org/10.2766/224310

Bonazza A, Sardella A, Cacciotti R, Kaiser A, Hanus C, De Nuntiis P, Hanus C, Maxwell I, Drdacký T, Drdacký M (2021) Safeguarding cultural heritage from climate change related hydrometeorological hazards in Central Europe. Int J Disaster Risk Reduct 63:102455, ISSN 2212-4209. https://doi.org/10.1016/j.ijdrr.2021.102455

Borrego S, Lavin P, Perdomo I, Gómez de Saravia S, Guiamet P (2012) Determination of indoor air quality in archives and biodeterioration of the documentary heritage. ISRN Microbiol 2012:1–10

Brimblecombe P (1990) The composition of museum atmosphere. Atmos Environ 24B:1–8. https://doi.org/10.1016/0957-1272(90)90003-D

Cacciotti R, Kaiser A, Sardella A, De Nuntiis P, Drdacky M (2021) Climate-change-induced disasters and cultural heritage: optimizing management strategies in Central Europe. Clim Risk Manag 32(15):100301. https://doi.org/10.1016/j.crm.2021.100301

Camuffo D (2007) Church heating and the preservation of the cultural heritage: guide to the analysis of the pros and cons of various heating systems. Electa Mondadori, Milan. ISBN: 88-370-5034-8

Camuffo D (2019) Microclimate for cultural heritage – measurement, risk assessment, conservation, restoration, and maintenance of indoor and outdoor monuments, 3rd edn, vol 23. Elsevier Science, New York, p 582. ISBN: 978–0–444-64106-9

Camuffo D, Sturaro G, Valentino A, Camuffo M (1999) The conservation of artworks and hot air heating systems in Churches: are they compatible? The case of Rocca Pietore. Italian Alps Stud Conserv 44(3):209–216. https://doi.org/10.2307/1506706

Caneva G, De Nuntiis P, Fornaciari M, Ruga L, Valenti P, Pasquariello G (2020) Aerobiology applied to the preventive conservation of cultural heritage. Aerobiologia 36:99–103. https://doi.org/10.1007/s10453-019-09589-9

Cappitelli F, Zanardini E, Sorlini C (2004) The biodeterioration of synthetic resins used in conservation. Macromol Biosci 4:399–406. https://doi.org/10.1002/mabi.200300055

Cassar M (1995) Environmental management: guidelines for museums and galleries. Routledge, London, p 165

Castillo NI, Ibáñez M, Beltrán E, Rivera-Monroy J, Ochoa JC, Páez-Castillo M, Posada-Buitrago ML, Sulyok M, Hernández F (2016) Identification of mycotoxins by UHPLC–QTOF MS in airborne fungi and fungi isolated from industrial paper and antique documents from the Archive of Bogotá. Environ Res 144(A):130–138. https://doi.org/10.1016/j.envres.2015.10.031

Cavallini T, Massa S, Russo A (1991) Optimal environmental conditions in museums. In: Science, technology and European cultural heritage: proceedings of the European symposium, Bologna, 13–16 June 1989. Butterworth-Heinemann Publishers, Oxford/Boston, pp 626–631. ISBN: 9780750602372

Ciferri O (1999) Microbial degradation of paintings. Appl Environ Microbiol 65(3):879–885. https://doi.org/10.1128/aem.65.3.879-885.1999

Cox CS, Wathes CM (1995) Bioaerosols handbook. CRC Press, Lewis Publishers, Boca Raton, p 656

De Guichen G (1980) Climate in museums: measurement. ICCROM, Rome

De Guichen G, Kabaoglu C (1985) How to make a rotten show-case. Museum (UNESCO) 146:64–67

De Nuntiis P, Maggi O, Mandrioli P, Ranalli G, Sorlini C (2003) Monitoring the biological aerosol. In: Mandrioli P, Caneva G, Sabbioni C (eds) Cultural heritage and aerobiology – methods and measurement techniques for biodeterioration monitoring. Kluwer Academic Publishers, Dordrecht, pp 107–144

Després VR, Huffman JA, Burrows SM, Hoose C, Safatov AS, Buryak G, Frohlich-Nowoisky J, Elbert W, Andreae MO, Pöschl U, Jaenicke R (2012) Primary biological aerosol particles in the atmosphere: a review. Tellus Ser B 64:15598. https://doi.org/10.3402/tellusb.v64i0.15598

Di Carlo E, Chisesi R, Barresi G, Barbaro S, Lombardo G, Rotolo V, Sebastianelli M, Travagliato G, Palla F (2016) Fungi and bacteria in indoor cultural heritage environments: microbial related risks for artworks and human health. Environ Ecol Res 4(5):257–264. https://doi.org/10.13189/eer.2016.040504

Douwes J, Thorne P, Pearce N, Heederik D (2003) Review bioaerosol health effects and exposure assessment: progress and prospects. Ann Occup Hyg 47:187–200. https://doi.org/10.1073/pnas.0811003106

Fröhlich-Nowoisky J, Pickersgill DA, Després VR, Poschl U (2009) High diversity of fungi in air particulate matter. Proc Natl Acad Sci 106:12814–12819. https://doi.org/10.1073/pnas.0811003106

Fuzzi S, Mandrioli P, Perfetto A (1997) Fog droplets – an atmospheric source of secondary biological aerosol particles. Atmos Environ 31:287–290. https://doi.org/10.1016/1352-2310(96)00160-4

Fuzzi S, Baltensperger U, Carslaw K, Decesari S, Denier van der Gon H, Facchini MC, Fowler D, Koren I, Langford B, Lohmann U, Nemitz E, Pandis S, Riipinen I, Rudich Y, Schaap M, Slowik JG, Spracklen DV, Vignati E, Wild M, Williams M, Gilardoni G (2015) Particulate matter, air quality and climate: lessons learned and future needs. Atmos Chem Phys 15:8217–8299. https://doi.org/10.5194/acp-15-8217-2015

Gaüzère C, Moletta-Denat M, Bousta F, Moularat S, Orial G, Ritoux S, Godon JJ, Robine E (2013) Reliable procedure for molecular analysis of airborne microflora in three indoor environments: an office and two different museum contexts. Clean–Soil Air Water 41(3):226–234. https://doi.org/10.1002/clen.201100699

Gaüzère C, Moletta-Denat M, Blanquart H, Ferreira S, Moularat S, Godon J-J, Robine E (2014) Stability of airborne microbes in the louvre museum over time. Indoor Air 24:29–40. https://doi.org/10.1111/ina.12053

Getty Conservation Institute (1994) Preventive conservation. In: Knell S (ed) Care of collections. Routledge, London, pp 83–87. https://www.taylorfrancis.com/chapters/edit/10.4324/9780203974711-18/preventive-conservation?context=ubx&refId=52f7f73c-7f59-4359-9bbd-9ee804d366aa

Gregory PH (1973) The microbiology of the atmosphere, 2nd edn. Aylesbury, Leonard Hall, p 377

Griffith WD, De Cosemo GAL (1994) The assessment of bioaerosols: a critical review. J Aerosol Sci 25(8):1425–1458. https://doi.org/10.1016/0021-8502(94)90218-6

Gudmundsson G, Monick MM, Hunninghake GW (1999) Viral infection modulates expression of hypersensitivity pneumonitis. J Immunol 162(12):7397–7401

Hinds WC (1999) Aerosol technology: properties, behavior, and measurement of airborne particles, 2nd edn. Wiley Interscience, New York, p 504

Hirst JM (1995) Bioaerosols: introduction, retrospect and prospect. In: Cox CS, Wathes CM (eds) Bioaerosols handbook. CRC Press, Lewis Publishers, Boca Raton, pp 5–14

Hueck HJ (2001) The biodeterioration of materials: an appraisal (reprinted). Int Biodeterior Biodegradation 48(1–4):5–11. https://doi.org/10.1016/S0964-8305(01)00061-0

IBC (ed) (2007) Oggetti nel tempo. Principi e tecniche di conservazione preventiva. Clueb, Bologna, pp 258. ISBN: 978-88-491-2866-6

Jaenicke R (2005) Abundance of cellular material and proteins in the atmosphere. Science 308(5718):73. https://doi.org/10.1126/science.1106335

Jones AM, Harrison RM (2004) The effects of meteorological factors on atmospheric bioaerosol concentrations, a review. Sci Total Environ 326(1–3):51–180. https://doi.org/10.1016/j.citotenv.2003.11.021

Kavklera K, Gunde-Cimermanb N, Zalarb P, Demšard A (2015) Fungal contamination of textile objects preserved in Slovene museums and religious institutions. Intern Biodeter Biodegr 97: 51–59. https://doi.org/10.1016/j.ibiod.2014.09.020

Kellogg CA, Griffin DW (2006) Aerobiology and the global transport of desert dust. Trends Ecol Evol 21(11):638–644. https://doi.org/10.1016/j.tree.2006.07.004

Lavin P, Gómez de Saravia SG, Guiamet PS (2014) An environmental assessment of biodeterioration in document repositories. Biofouling 30(5):561–569. https://doi.org/10.1080/08927014.2014.897334

Lazaridis M, Katsivela E, Kopanakis I, Raisi PG (2015) Indoor/outdoor particulate matter concentrations and microbial load in cultural heritage collections. Herit Sci 3:34. https://doi.org/10.1186/s40494-015-0063-0

Letch T (2016) Evaluation of a parchment document, the 13th century incorporation charter for the city of Krakow, Poland, for microbial hazards. Appl Environ Microbiol 82:2620–2631. https://doi.org/10.1128/AEM.03851-15

Lighthart B, Stetzenbach LD (1994) Distribution of microbial bioaerosol. In: Lighthart B, Mohr AJ (eds) Atmospheric microbial aerosols: theory and applications. Springer, Boston, MA, pp 68–98. https://doi.org/10.1007/978-1-4684-6438-2_4

MacNeil L, Kauri T, Robertson W (1995) Molecular techniques and their potential application in monitoring the microbiological quality of indoor air. Can J Microbiol 41(8):657–665. https://doi.org/10.1139/m95-091

Mandrioli P (1985) L'Aerobiologia e le sue moderne applicazioni. Aerobiologia 1:2–4. https://doi.org/10.1007/BF02447689

Mandrioli P (2015) The importance of environmental parameters for the conservation of mural paintings. Specificity of the Sistine environment. Conference proceedings: the Sistine Chapel, twenty years later. New breath new light, Auditorium Conciliazione, Roma 30-31/10/2014. Edizioni Musei Vaticani, Roma, pp 119–133. ISBN: 978-88-8271-366-9

Mandrioli P, Ariatti A (2001) Aerobiology: future course of action. Aerobiologia 17:1–10. https://doi.org/10.1023/A:1007602928928

Mandrioli P, Negrini MG, Scarani C, Tampieri F, Trombetti F (1980) Mesoscale transport of Corylus pollen grains in winter atmosphere. Grana 19(3):227–233. https://doi.org/10.1080/00173138009425007

Mandrioli P, Negrini MG, Cesari G, Morgan G (1984) Evidence for long range transport of biological and anthropogenic aerosol particles in the atmosphere. Grana 23(1):43–53. https://doi.org/10.1080/00173138409428876

Mandrioli P, Comtois P, Levizzani V (eds) (1998) Methods in aerobiology. Pitagora Editrice, Bologna, pp. 276. ISBN: 88-371-1043-X

Michalski S (1994) A systematic approach to preservation: description and integration with other museum activities. Stud Conserv 39(S-2):8–11

Micheluz A, Manente S, Tigini V, Prigione V, Pinzari F, Ravagnan G, Varese GC (2015) The extreme environment of a library: xerophilic fungi inhabiting indoor niches. Int Biodeterior Biodegradation 99:1–7. https://doi.org/10.1016/j.ibiod.2014.12.012

Montacutelli R, Maggi O, Tarsitani G, Gabrielli N (2000) Aerobiological monitoring of the "Sistine Chapel": airborne bacteria and microfungi trends. Aerobiologia 3:441–448. https://doi.org/10.1023/A:1026525432412

Năşcuţiu AM (2010) Viable non-culturable bacteria. Bacteriol Virusol Parazitol Epidemiol 55:11–18

Nugari MP (2003) The aerobiology applied to the conservation of works of art. In: Saiz-Jimenez C (ed) Coalition advanced course on biological problems in cultural heritage, Florence 8--9 November 2002. Coalition Newsletter 6(SI): 8–10

Nugari MP, Ricci S, Roccardi A, Monte M (2003) Churches and hypogea. In: Mandrioli P, Caneva G, Sabbioni C (eds) Cultural heritage and aerobiology – methods and measurement techniques for biodeterioration monitoring. Kluwer Academic Publishers, Dordrecht, pp 207–224

Ortega-Morales BO, Narváez-Zapata J, Reyes-Estebanez M, Quintana P, De la Rosa SDC (2016) Bioweathering potential of cultivable fungi associated with semi-arid surface microhabitats of Mayan buildings. Front Microbiol 7:201. https://doi.org/10.3389/fmicb.2016.00201

Paiva de Carvalho HP, Sequeira SO, Pinho D, Trovao J, Fernandes da Costa RM, Egas C, Macedo MF, Portugal A (2019) Combining an innovative non-invasive sampling method and high-throughput sequencing to characterize fungal communities on a canvas painting. Int Biodeterior Biodegradation 145:104816. https://doi.org/10.1016/j.ibiod.2019.104816

Palla F (2015) Biological risk for the conservation and exhibition of historical–artistic artefacts in confined spaces. In: Lorusso S, Natali A, Palla F (eds) Risk management in the cultural heritage sector: museums, libraries and archives. Mimesis Edizioni, Milano-Udine, pp 95–108

Palla F, Di Carlo E, Barresi G, Billeci N, Chisesi R, Barbaro S, Sebastianelli S (2014) Bioaerosol in indoor environments for conservation of cultural heritage: potential risks for historical-artistic manifacts and human health. In: International Conference Indoor Air Quality in Heritage and Historic Environments, Prague, Kaiserštejn Palace, April 13–16 2014, p 47

Pashley CH, Fairs A, Free RC, Wardlaw AJ (2012) DNA analysis of outdoor air reveals a high degree of fungal diversity, temporal variability, and genera not seen by spore morphology. Fungal Biol 116(2):214–224. https://doi.org/10.1016/j.funbio.2011.11.004

Pasquarella C, Pitzurra O, Savino A (2000) The index of microbial air contamination. J Hosp Infect 46:241–256. https://doi.org/10.1053/jhin.2000.0820

Pasquarella C, Balocco C, Pasquariello G, Petrone G, Saccani E, Manotti P, Ugolotti M, Palla F, Maggi O, Albertini R (2015) A multidisciplinary approach to the study of cultural heritage environments: experience at the Palatina library in Parma. Sci Total Environ 536:557–567. https://doi.org/10.1016/j.scitotenv.2015.07.105

Pitzurra M, Savino A, Pasquarella C, Poletti L (1997) A new method to study the microbial contamination on surfaces. Hyg Med 22:77–92

Pöschl U (2005) Atmospheric aerosols: composition, transformation, climate and health effects. Angew Chem Int Ed Engl 44(46):7520–7540. https://doi.org/10.1002/anie.200501122

Puškárová A, Bučková M, Habalová B, Kraková L, Maková A, Pangallo D (2016) Microbial communities affecting albumen photography heritage: a methodological survey. Sci Rep 6(20810). https://doi.org/10.1038/srep20810

Ramos T, Stephens B (2014) Tools to improve built environment data collection for indoor microbial ecology investigations. Build Environ 81:243–257. https://doi.org/10.1016/j.buildenv.2014.07.004

Ranalli G, Zanardini E, Andreotti A, Colombini MP, Corti C, Bosch-Roig P, De Nuntiis P, Lustrato G, Mandrioli P, Rampazzi L, Giantomassi C, Zari D (2018) Hi-tech restoration by two steps biocleaning process of triumph of death fresco at the Camposanto monumental cemetery (Pisa, Italy). J Appl Microbiol 125(3):800–812. https://doi.org/10.1111/jam.13913

Rantio-Lehtimäki A (1994) Short, medium and long range transported airborne particles in viability and antigenicity analyses. Aerobiologia 10(2):175–181. https://doi.org/10.1007/BF02459233

Ruga L, Orlandi F, Romano B, Fornaciari M (2015) The assessment of fungal bioaerosols in the crypt of St. Peter in Perugia (Italy). Int Biodeter Biodegradr 98:121–130. https://doi.org/10.1016/j.ibiod.2014.12.010

Sabbioni C, Mandrioli P, Bonazza A, De Nuntiis P, Guaraldi C (2008) Indagini diagnostiche finalizzate alla identificazione delle cause di degrado degli affreschi del camposanto monumentale di Pisa. Report tecnico contratto OPA Pisa – ISAC CNR, Bologna, p 154

Saiz-Jimenez C, Gonzalez JM (2007) Aerobiology and cultural heritage: some reflections and future challenges. Aerobiologia 23:89–90. https://doi.org/10.1007/s10453-007-9059-x

Sanmartín P, De Araujo A, Vasanthakumar A (2016) Melding the old with the new: trends in methods used to identify, monitor, and control microorganism on cultural heritage materials. Microb Ecol 71(4):1–17. https://doi.org/10.1007/s00248-016-0770-4

Schlesinger P, Mamane Y, Grishkan I (2006) Transport of microorganisms to Israel during Saharan dust events. Aerobiologia 22:259–273. https://doi.org/10.1007/s10453-006-9038-7

Schwendemann AB, Wang G, Osborn JM, Thatcher SL (2007) Aerodynamics of saccate pollen and its implications for wind pollination. Am J Bot 94(8):1371–1381. https://doi.org/10.3732/ajb.94.8.1371

Sedlbauer K (2002) Prediction of mould growth by hygrothermal calculation. J Build Phys 25:321–336. https://doi.org/10.1177/0075424202025004093

Sesana E, Gagnon AS, Ciantelli C, Cassar J, Hughes JJ (2021) Climate change impacts on cultural heritage: a literature review. Wiley Interdiscip Rev-Clim Chang 12(e710). https://doi.org/10.1002/wcc.710

Šimonovičová A, Kraková L, Pangallo D, Majorošová M, Piecková E, Bodoriková S, Dörnhoferová M (2015) Fungi on mummified human remains and in the indoor air in the Kuffner family crypt in Sládkovičovo (Slovakia). Int Biodeter Biodegrad 99:157–164. https://doi.org/10.1016/j.ibiod.2014.12.011

Skóra J, Gutarowska B, Pielech-Przybylska K, Stępień Ł, Pietrzak K, Piotrowska M, Pietrowski P (2015) Assessment of microbiological contamination in the work environments of museums, archives and libraries. Aerobiologia 31(3):389–401

Sterflinger K (2010) Fungi: their role in the deterioration of cultural heritage. Fung Biol Rev 24:47–55. https://doi.org/10.1016/j.fbr.2010.03.003

Sterflinger K, Pinzari F (2012) The revenge of time: fungal deterioration of cultural heritage with particular reference to books, paper and parchment. Environ Microbiol 14(3):559–566

Su C, Lei L, Duan Y, Zhang KQ, Yang J (2012) Culture-independent methods for studying environmental microorganisms: methods, application, and perspective. Appl Microbiol Biotechnol 93(3):993–1003. https://doi.org/10.1007/s00253-011-3800-7

Tampieri F, Mandrioli P, Puppi GL (1977) Medium range transport of airborne pollen. Agr Meteorol 18(1):9–20. https://doi.org/10.1016/0002-1571(77)90023-1

Tarsitani G, Moroni C, Cappitelli F, Pasquariello G, Maggi O (2014) Microbiological analysis of surfaces of Leonardo Da Vinci's Atlantic codex: biodeterioration risk. Int J Microbiol 2014: 214364–214371. https://doi.org/10.1155/2014/214364

Thomson G (1986) The museum environment, 2nd edn. Butterworths, Los Angeles, p 292

United Nations, Sendai Framework for Disaster Risk Reduction 2015–2030 (2015) United Nations Office for Disaster Risk Reduction (UNISRD)

Urzì C, De Leo F, De Hoog GS, Sterflinger K (2000) Recent advances in the molecular biology and ecophysiology of meristematic stone-inhabiting fungi. In: Ciferri O, Tiano P, Mastromei G (eds) Of microbes and art. The role of microbial communities in the degradation and protection of cultural heritage. Kluwer Academic Plenum, London, pp 3–19

Valentin N (2003) Microbial contamination in museums collections: organic materials. In: Saiz-Jimenez C (ed) Molecular biology and cultural heritage. Balkema Pub, Lisse, pp 85–91

Varas-Muriel MJ, Fort R, Martínez-Garrido MI, Zornoza-Indart A, López-Arce P (2014) Fluctuations in the indoor environment in Spanish rural churches and their effects on heritage conservation: hygro-thermal and CO_2 conditions monitoring. Build Environ 82:97–109. https://doi.org/10.1016/j.buildenv.2014.08.010

Walser SM, Gerstner DG, Brenner B, Bünger J, Eikmann T, Janssen B, Kolb S, Kolk A, Nowak D, Raulf M, Sagunski H, Sedlmaier N, Suchenwirth R, Wiesmüller G, Wollin K-M, Tesseraux I, Herr C (2015) Evaluation of exposure-response relationships for health effects of microbial bioaerosols—a systematic review. Int J Hyg Environ Health. https://doi.org/10.1016/j.ijheh.2015.07.004

3 Biocide

Giovanna Barresi, Maria Giovanna Parisi, Valentina Rotolo,
Matteo Cammarata, and Franco Palla

Abstract

Biodeterioration represents a revealing problem for the conservation of cultural heritage. It can be identified as a complex interaction within the ecosystem of a microbial community and its substrate and involves physical and chemical alterations resulting from biological and metabolic activity. Designing a diagnostic approach for evaluating the extent of the damage, identifying the biological community, and opting for an efficient methodology aimed at eliminating deteriogens is equally complicated. The correct approach would require understanding the nature of the biodeterioration and implementing methodologies respectful of human health which, however, avoid the indiscriminate killing of organisms. Different preventive or remedial methods are used for this purpose. They include well-known physical and mechanical methods with their operating limitations and the most frequently used chemical methods, supported by biocide products for the elimination or growth inhibition of target organisms. Unfortunately, most—if not all—biocides applied on artworks are toxic or otherwise polluting substances, and their degradation is frequently difficult, being persistent in the natural environment. Moreover, due to the fact that there are no specific formulations destined for conservation practice, commercial biocide products come from the medical or agricultural field, carrying with them their well-known negative effects. Research in this sector focuses on ways to replace toxic products with natural molecules that do not cause adverse effects, in

G. Barresi (✉) · V. Rotolo · F. Palla
Department of STEBICEF, Laboratory of Biology and Biotechnologies for Cultural Heritage, University of Palermo, Palermo, Italy
e-mail: franco.palla@unipa.it

M. G. Parisi · M. Cammarata
Department DiSTeM, Laboratory of Immunobiology, University of Palermo, Palermo, Italy
e-mail: mariagiovanna.parisi@unipa.it; matteo.cammarata@unipa.it

F. Palla, G. Barresi (eds.), *Biotechnology and Conservation of Cultural Heritage*,
https://doi.org/10.1007/978-3-030-97585-2_3

addition to the application of alternative methods and the support of formulations for safe nontoxic novel compounds.

Keywords

Microbial colonization · Minimum biocidal concentration · Essential oils · Hydroalcoholic extract · in vitro analysis

3.1 Prevention and Control of Microbial Colonization

When biological agents are detrimental or cause aesthetic damage on materials composing cultural heritage, several conservative treatments and products are used, in particular, to reduce or eliminate biological colonization (Chapman 2003; Caneva et al. 2008). The first step is to examine the physical and chemical factors that have facilitated the initial colonization and its expansion on the cultural objects and monuments and to establish which mechanisms and regulatory processes are involved (Cappitelli et al. 2011; Warscheid 2000, 2003; Negi and Sarethy 2019), defining ad hoc solutions in relation to the artworks and their environments (e.g., museums, outdoor heritage, archives, archaeological sites). Furthermore, each procedure should be carefully planned to guarantee the integrity of the artifacts, ensuring that its effect remains active for as long as possible by preventing recolonization and that no hazards exist for operators or the environment (Caneva et al. 1994).

Conservation is typically based on the use of *preventive* and *remedial* methods that inhibit or eradicate microbial growth and differ according to the nature of the artifact, its conservation state, dimensions, and location and must also take into consideration environmental conditions and other variables (Sterflinger and Pinzari 2012; Orlita 2004; Young et al. 2008).

Preventive (indirect) methods are based on the monitoring and setting of parameters and factors contributing to microbial development, such as temperature, pH, and moisture content/relative humidity. Remedial (direct) methods involve the use of mechanical, physical, chemical (biocides, insecticides, pesticides), and biochemical measures (Sterflinger and Pinzari 2012; Kumar and Kumar 1999; Caneva et al. 2008; Koestler and Santoro 1990).

Preventive methods include all activities aimed at inhibiting biological attack, by modifying, where possible, environmental conditions and physicochemical parameters on a surface, so they become unfavorable for biological growth without acting directly on the material constituting the artwork. However, environmental factors cannot always be controlled, whereas it is possible to control indoor environmental conditions (e.g., museums, libraries, archives, churches). It is decidedly more difficult to control the external environmental conditions (outdoor monuments and archaeological sites) (Rives and Talegon 2006; Stupar et al. 2014); on the contrary, factors encouraging biodeterioration (dirt, dust, and deposits of a different kind) can be removed. Effectively, preventive measures also include routine maintenance, periodic cleaning of dirt, and removal of dust—which contains a high level of microbial contamination and is especially rich in bacterial and fungal species, which may cause biodeterioration and may represent a risk for artwork conservation—in addition, water repellent and consolidating treatments can aim at inhibiting biological attack. Unfortunately, these preventive measures are not always correctly followed (Kumar and Kumar 1999; Doehne and Price 2010; Sterflinger and Pinzari 2012; Michaelsen et al. 2013; Fernandes 2006; Vukojević and Grbić 2010). In the case of documentary heritage, the main way to avoid degradation is preservation; climate control, guaranteeing some conditions for optimal temperatures (15–25 °C), pH, illumination, ventilation, relative humidity (40–65%); frequent cleaning; monitoring of the objects (Sterflinger and Pinzari 2012; Roman et al. 2013). The protection of paper from damage by microorganisms mainly consists in inhibiting fungal activity; this can be achieved by various means with the use of various biocides. The problem is generally in selecting the most appropriate biocide and method of treatment (Velikova et al. 2011).

Control measures, or remedial methods, provide for direct action on the biodeteriogens with interventions on artwork surfaces and performing mechanical, physical, biological, chemical, or biochemical actions.

Mechanical methods have the disadvantage of failing to guarantee results over the long term; where possible, they are used in combination with chemical procedures and can thus be very useful (Caneva et al. 1994).

Physical methods of disinfection consist in the use of radiation with biocide action, X-rays, gamma, and UV-C radiations (Pointing et al. 1998; Sterflinger and Pinzari 2012; Michaelsen et al. 2013; Sequeira et al. 2012; Borderie et al. 2014). Gamma radiation is very effective against fungi and their spores, but its application is restricted, as it causes depolymerization of cellulose in paper-based materials, together with aging and recolonization (Magaudda 2004; Adamo et al. 2003).

Chemical methods suitable for use in cultural heritage consist in a wide range of products, mostly based on organic and inorganic compounds, quaternary ammonium compounds, phenol and urea derivatives, and nitrogen-containing compounds (Fernandes 2006). Considering the heterogeneity of chemicals used as biocides, they may have more than one potential target, unlike antibiotics that have very specific targets. The chemical structure of a biocide determines its affinity to specific targets and is the key to understanding its mode of action. The spectrum of activity is determined against bacteria (Gram-positive, Gram-negative, mycobacteria, etc.) and fungi (mold and yeast) and dose–response relationships are established. There are numerous methods to evaluate biocide efficiency, including a basic microbiological simulation and field tests to demonstrate the effectiveness of a biocide (Sondossi 2009; Ranalli et al. 2003; Cappitelli et al. 2009; Speranza et al. 2012). A small number of biocide agents have been tested with respect to their compatibility with historical-artistic materials, and only few studies have shown the long-term effects of biocides (Koestler et al. 1993; Ascaso et al. 2002; Tretiach et al. 2007). Besides compatibility with the materials of the treated artifacts, in many cases, the most challenging aspect of biocide treatments is that objects are infested by a mixed community of microorganisms with different levels of susceptibility toward the

Table 3.1 Chemical and commercial biocides used on different substrates

Biocide	Target organism	Substratum	References
Koretrel	Epilithic photoautotrophic organisms	Limestone, sandstone	Tretiach et al. (2007)
Neo Desogen Metatin 5810–101	Cyanobacteria	Frescoes	Cappitelli et al. (2009)
Anatase Biotin T Anios	Cyanobacteria and chlorophyta species	Mortars	Fonseca et al. (2010)
Sanatex Anti-B Rocima GT Rocima 243 Phosphopag	Micromycetes	Paper	Velikova et al. (2011)
Koretrel	Lichens	Stone	Speranza et al. (2012)
Biotin T Algophase Preventol	Micromycetes	Frescoes	Maxim et al. (2012)

chemical compound applied (Chapman 2003; Bastian and Alabouvette 2009; Martin-Sanchez et al. 2012; Fernandes 2006).

Chemical treatments include liquid biocides and fumigation using very toxic gases, such as methyl bromide, sulfuryl fluoride, and ethylene oxide (Sterflinger and Pinzari 2012; Michaelsen et al. 2013; Sequeira et al. 2012; Velikova et al. 2011). European conservators have used fumigation because it does not affect cellulosic material. This type of treatment, however, needs special vacuum chambers and must be executed by specialists due to the toxicity of ethylene oxide and the strict regulations regarding its use (Borrego et al. 2012; Rakotonirainy and Lavédrine 2005).

The most common biocides in the field of restoration have as their active ingredients benzalkonium chloride, permethrin, nystatin (antibiotics), camphor, thymol, sodium fluoride, naphthalene, ethylene oxide (gas), nitrogen (gas), and other compounds applied for the disinfection of fungi and insects. These chemicals are applied on polychrome artifacts, wooden sculptures, paintings and panels, textiles, and paper artifacts. Even though commercial biocides have been widely applied on different cultural materials (Table 3.1), insufficient studies have been carried out on the efficiency of the products over time and on the adverse effects for cultural heritage and humans. Recently, trends in biodeterioration control indicate the need to include biocide procedures using nonharmful and nontoxic products in order to control or to eradicate different types of microbial colonization.

3.2 Risks for Humans and the Environment

When considering biocides to control and eradicate biological colonization, several factors, such as toxicity to humans, risks of environmental pollution and compatibility with substrates, need to be discussed. Many biocides are difficult to degrade and are persistent in environment, also causing contamination of areas far from the site of treatment (Guiamet et al. 2006; Marcotte et al. 2014). Considerable information can be found in literature, particularly in the field of agriculture, ecology, hydrology, health, and environmental science. The risk is linked to undesirable effects in nontarget microflora, plants, animals, and aquatic life—especially relevant where the risk of soil and water contamination is high (Fleeger et al. 2003; Blanchoud et al. 2004; Arias-Estévez et al. 2008), since large quantities of biocides are released into the environment and accumulate, resulting in a variety of negative effects (Marshall and McMurry 2005; Schieweck et al. 2007; Hernández et al. 2017). To protect nontarget species from damage due to direct or indirect exposure to these toxic substances, federal environmental protection agencies (FEPA) have established specific standards for their use, handling, and disposal (Damalas and Eleftherohorinos 2011).

Biocide toxicity is generally quantified by using LD50 (lethal dose 50%) and LC50 (lethal concentration 50%) values as indices. LD50 is the amount of active substance that can be expected to cause death in half (50%) of experimental animal species exposed. LC50 is usually reported as milligrams of a substance per cubic meter of the atmosphere to which the animal is exposed over a set time (Kumar and Kumar 1999; Damalas and Eleftherohorinos 2011). To reduce health hazards associated with the use of biocides, specific control measures and work practices are recommended, such as good ventilation, personal respiratory protection, and safety equipment (Ströfer-Hua 1993). Tomei et al. (1996) have shown the effects of solvents in the blood of art restorers, related to restoration work. It may be necessary to assess toxic compound exposure in those environments where biocide substances are used, in order to define safe levels when evaluating risk (Guiamet and Gómez de Saravia 2005; Hahn et al. 2010).

It should be kept in mind that biocides are all toxic by definition and most are also corrosive. Recognition for the need to control risks from biocides has come from scientists in academia, regulatory agencies, and industry. Hazard identification of a substance and the identification of related adverse effects are likened to the dose or level of exposure to humans, animals, and the environment (Paulus 2005; Sondossi 2009).

Recently, many researchers have shown the possibility of using new methodologies for eradicating microbial colonization on cultural heritage without the use of harmful substances.

3.3 Novel Molecules with Antimicrobial Activity

Many strategies have been developed and tested as an alternative to traditional biocides (Silva et al. 2015; Sasso et al. 2013; Guiamet et al. 2006; Sequeira et al. 2012; Mansour and Ahmed 2012). These innovative approaches include the use of synthetic or natural molecules with antimicrobial activity.

3.3.1 Synthetic Molecules

Excellent results have been obtained in medical and pharmaceutical fields by using metal nanoparticles (e.g., silver, copper, platinum, gold, titanium, zinc) in antimicrobial coatings (Rai et al. 2010; Knetsch and Koole 2011; Fu et al. 2005). The antibacterial properties of silver nanoparticles (AgNPs) have been investigated to control microbial colonization on the surface of historical objects, archival documents, and stone materials (Bellissima et al. 2014; Gutarowska et al. 2012, 2014; Essa and Khallaf 2014; Carrillo-González et al. 2016; Kumar et al. 2008). The antimicrobial effect of AgNPs against microbial genera (eight bacteria and seven fungi) isolated from archaeological mural biofilms has been tested (Carrillo-González et al. 2016); these works showed several properties for nanoparticles, such as thermal stability and low toxicity.

Moreover, some researchers have focused their attention on biocide efficiency of zinc nanopowders (ZnO and $ZnTiO_3$) and antimicrobial photocatalytic coatings (TiO_2) (Ruffolo et al. 2010; La Russa et al. 2012). Nanoparticles of ZnO or TiO_2, added to suspensions of $Ca(OH)_2$, were applied on two limestone lithotype surfaces; both coatings displayed antifungal activity against *Penicillium oxalicum* and *Aspergillus niger* (Gómez-Ortíz et al. 2013). Two silsesquioxane-based hybrid nanocomposites (methacrylate units containing titania and/or silver nanoparticles) utilized for coatings of monumental stones showed high antibacterial/antifungal efficiency (Aflori et al. 2013). Zinc oxide nanoparticles (ZnO-NPs) were tested on stone specimens evaluating the possible chromatic change after treatment, and the biostatic activity of bioactive nanocomposites was assessed against *Aspergillus niger* (Ditaranto et al. 2015; van der Werf et al. 2015). Recently, novel compounds have been tested for the conservation of wood or used in underwater archaeological sites (Oliva et al. 2015; Walsh et al. 2016; Ruffolo et al. 2013).

3.3.2 Natural Molecules from Animals

Recently, several publications have highlighted that some natural products (i.e., from animal and plants) may represent sustainable alternatives to traditional procedures to control microbial colonization (Valgas et al. 2007; Barresi et al. 2015; de la Paz et al. 2006; Borrego et al. 2012; Casiglia et al. 2014, Anand et al. 2019).

Antimicrobial peptides (AMPs) are typically small molecules (up to 3000 Da) naturally produced in cells and tissues of invertebrate, plant, and animal species, so they are ubiquitous for their importance in building defense strategies (Brogden 2005). In Humans, these peptides are mainly found in those tissues that come into contact with environmental pathogens (skin, ear, eyes, and epithelial surface) (Wang 2014). AMPs have been demonstrated to have their own advantages over the traditional antibiotics with a broad-spectrum of antimicrobial activities including anti-bacteria, anti-fungi, anti-viruses, and anti-cancers, and even overcome bacterial drug resistance (Hancock and Scott 2000; Lei et al. 2019). Moreover, AMPs are considered part of the humoral natural defense of invertebrates against infections and have thus been termed "natural antibiotics." In invertebrates, antimicrobial peptides represent the major humoral defense system against infection, showing a diverse spectrum of action mechanisms, most of them related to plasma membrane disturbance and lethal alteration of microbial integrity (Otero-González et al. 2010). Marine invertebrates, especially sedentary sea anemones are evolved with rich sources of bioactive metabolites, which could be used for novel antimicrobial drugs (Hamayeli et al. 2019; Mohammadi et al. 2019; Trapani et al. 2015; Donia and Hamann 2003; Williams et al. 2007; Pomponi 1999; Jimeno 2002; Gordon et al. 1994; Carté 1996).

The study of biologically active compounds produced by marine organisms is rooted in the discovery of unusual nucleoside derivatives in extracts from *Tethya crypta*, a sponge designed by Bergmann and Feeney in 1950 (Bergmann and Feeney 1950, 1951). In the late 1960s, for the first time, the prostaglandin from a species of gorgonian, typical of the Caribbean Sea, *Plexaura homomalla*, was isolated (Proksch et al. 2002; Weinheimer and Spraggins 1969); up to 2010, more than 15,000 natural substances produced by marine invertebrates were isolated and tested (Kinghorn 2008). Since the early 1950s over 20,000 MNPs have been discovered, including compounds produced by symbiotic bacteria, referred to as "blue gold" in the search for new drugs (Blockley et al. 2017), with over 8000 compounds registered only in the decade between 2001 and 2010 (Faulkner 2002; Blunt et al. 2006; Datta et al. 2015; Patra et al. 2020).

As reported by *Blue Biotechnology*, marine invertebrates and vertebrates (sponges, tunicates, jellyfish, sea anemones, shellfish, and fishes) represent an important resource for health, food and processing, and preservation industries (Destoumieux-Garzón et al. 2016; Salomone et al. 2012; Otero-González et al. 2010; Smith et al. 2010; Alonso et al. 2003). Some of these molecules are totally safe for humans and the environment and are both stable and active at low temperatures, performing a specific action throughout a short reaction time. As shown by Barresi et al. (2015), these bioactive molecules may represent a valid alternative to traditional procedures in sustainable restoration projects. Despite the enriched bioactive compounds, the marine invertebrates are largely unexplored as identification and application assessment of lead compounds and their synthetic analogs remain a major task to be solved.

3.3.3 Essential Oils

Plant-based biocides are generally hydroalcoholic extracts or essential oils (Acharya et al. 2011; Popovici et al. 2019) mostly used as biocides in medical, food, and pharmaceutical industries; it has been demonstrated that natural derivate compounds are highly efficient against bacteria, fungi, and yeasts (Dubey et al. 2000; Feng and Zheng 2007; Kumar et al. 2010; Walentowska and Foksowicz-Flaczyk 2013; Sakr et al. 2012). Biocide action is attributable to different secondary metabolites, including essential oils, triterpenoids, flavonoids, phenols, alkaloids, coumarins, tannins, and steroids, with different models of action, such as regulation of intermediate metabolism, activation or blocking of enzymatic reactions, and alteration of membrane structures (Kumar and Simon 2016; Guiamet et al. 2006; Kalemba and Kunicka 2003; Yang and Clausen 2007; Guiamet et al. 2008).

Essential oils from *Ocimum gratissimum* (basil), *Zingiber cassumunar* (ginger), *Cymbopogon citratus* (lemongrass) and *Caesulia axillaris* (caesulia) display high efficiency against *Aspergillus flavus*, as well as thyme, sage, nutmeg, and cassia against *Alternaria alternata* (Kumar et al. 2010; Dubey et al. 2000; Feng and Zheng 2007). Arctium and centaurea have been tested for their antimicrobial, antiseptic, and disinfectant actions, given by their content in coumarins, flavonoids, organic acids, and polyacetylene compounds (Gómez de Saravia et al. 2008; Ebadollahi et al. 2020). In addition, the antimicrobial activity of other essential oils extracted from anise seed, cloves, cumin, garlic, laurel, orange sweet, and oregano is well known (Borrego et al. 2012). Researchers from different countries have tested the antimicrobial activity of several plant extracts and essential oils, using native plants, against microorganisms associated with biodeterioration in archives, libraries, and museums (Borrego et al. 2012; Gómez de Saravia et al. 2008; Guiamet et al. 2008; Rotolo et al. 2016). Essential oils (lemon, spearmint, fennel, marjoram, rosemary) have been assayed at different concentrations against yeast colonies isolated from royal tombs (limestone and granite blocks) at Tanis (Ghalem and Mohamed 2008). The antimicrobial activity of *Ricinus communis* against different bacteria isolated from indoor air documents repositories in the National Archives of Cuba has been determined (de la Paz et al. 2006).

Guiamet et al. (2008) selected extracts from several species of plants (*Cichorium intybus* L., *Arctium lappa* L., *Centaurea cyanus* L., *Medicago sativa* L., *Plantago major* L., *Eucalyptus citriodora Hook*, *Pinus caribaea Morelet*, *Allium sativum* L., *Piper auritum Kunth*) estimating the minimum inhibitory concentration (MIC). The antimicrobial activity was tested against *Bacillus* sp., isolated from the surface of photographic paper stored in the Historical Archive in Argentina. Considerable interest in oregano (*Origanum vulgare* L., Lamiaceae) extracts and essential oil has been established in agricultural, pharmaceutical, and cosmetic industries (Şahin et al. 2004). The antifungal activity of *Origanum vulgare*, *Rosmarinus officinalis*, and *Lavandula angustifolia* (Lamiaceae) essential oils has been investigated against fungi (*Bipolaris spicifera* and *Epicoccum nigrum*) isolated from stone and wooden (*Aspergillus niger*, *Aspergillus ochraceus*, *Penicillium* sp., and *Trichoderma viride*) objects (Stupar et al. 2014). The antifungal activity of *Origanum vulgare* L. and

Thymus vulgaris L. has been evaluated by Lavin et al. (2016) against *Scopulariopsis* sp. and *Fusarium* sp. isolated from paper documents. *O.vulgare* essential oil has been also tested against even *Aspergillus* species isolated from different substrata (stone, brick, silk, and paper) of cultural heritage objects in Serbia by Savković et al. (2016), comparing the biocide activity of commercial biocides (silver ions and hydrogen peroxide).

Recently Palla et al. (2020) performed tests using *Origanum vulgare* or *Thymus vulgaris* essential oils (EOs) against fungal colonization or insect infestation exposing to the volatile compound. In the study, the activity of the EOs was investigated as a green biocide against *Aspergillus flavus* colonies isolated from the base of a wooden sculpture or as a green repellent of *Anobium punctatum* insects that had infested the heads of wooden Sicilian puppets, both artifacts are from the eighteenth century.

Casiglia et al. (2019) confirmed that thymus essential oil has good antimicrobial activity, comparable with the action of chloramphenicol and ketoconazole, against *Bacillus subtilis* and *Staphylococcus epidermidis*, besides excellent antifungal properties against *Fusarium oxysporum* and *Aspergillus niger* (microorganisms quite frequently infesting archives, libraries, and historical art craft objects). Moreover, several plant essential oils can be applied in the prevention of fungal infections, demonstrating an antifungal activity greater than some commercial fungicides (Soković et al. 2009), and due to their safety and selective action can be applied in the sustainable conservation of artworks (Vukojević and Grbić 2010). The antifungal activity of the essential oils (hydroalcoholic/water extracts) of *Ocimum basilicum* against *Aspergillus* sp., *Penicillium* sp., and *Mucor* sp. colonizing paper artifacts has also been investigated (Fierascu et al. 2014). The control of fungal colonization (*Fusarium oxysporum*, *Aspergillus niger*, *Alternaria alternata*) on Egyptian stucco ornaments has been performed by some plant extracts (*Anethum graveolens*, *Cymbopogon citrates*, *Juniperus oxycedrus*) or essential oils, including *Tea tree oil*, against the most cellulolytic fungi (*Penicillium* sp., *Aspergillus flavus*, *Trichoderma viride*, *Alternaria tenuis*, and *Aspergillus niger*) isolated from the aerosol of the National Library and Archives of Egypt (Afifi 2012; Sahab Ahmed et al. 2014). However, the antimicrobial effect of plant extracts in archives, libraries, and museums has still not been sufficiently explored (Guiamet et al. 2008), and further studies should be carried out to evaluate their successful application in biodeterioration control.

Currently, unlike antibiotics, a standardized protocol to evaluate the inhibitory activity of plant extracts is lacking, although the Agar disk diffusion or well diffusion are the used techniques (Al-Hussaini and Mahasneh 2009; Calvo et al. 2001; Ćirković et al. 2012; Valgas et al. 2007; Palla et al. 2021).

Plant extracts have also been applied to contrast insect pests in a gazelle mummy (Abdel-Maksoud et al. 2014) and in laboratory specimens, testing directly on termites (Bacci et al. 2015; Ebadollahi et al. 2020; Palla et al. 2020).

Salem et al. (2016) tested two oils for inhibiting fungal growth on wooden samples. The activity of the essential oils *Pinus rigida* (wood) and *Eucalyptus camaldulensis* (leaves) was estimated against five common mold fungi (*Alternaria alternata*, *Fusarium subglutinans*, *Chaetomium globosum*, *Aspergillus niger*, and *Trichoderma viride*). The findings support the potential use of the essential oils in wood protection for surface treatment or fumigation.

Moreover, fumigation by essential oils (geranium, spikenard, muskmelon, and patchouli) against museum insects has been investigated and can be considered an interesting alternative to ethylene oxide fumigation (Borrego et al. 2012; Chun et al. 2000).

Recently, in order to improve antimicrobial efficacy of *Thymus vulgaris* L. essential oil, two strategies have been performed. In the first, parchment artworks showing huge colonization by *Bacillus* sp., *Georgenia* sp., *Ornithinibacillus* sp. and *Streptomyces* sp., were exposed to EO volatile compounds in a dedicated plexiglass chamber under vacuum condition, that significantly reduced the time of exposure (D'Agostino et al. 2021). In the other, the potential synergic action of *T. vulgaris* essential oil and hydroalcoholic extract on wooden sculptures surface was evaluated. Specifically, before exposition to essential oil volatile compound (environmental condition) hydroalcoholic extract was directly applied on wooden sculpyure surface. This strategy turned out to be ideal for artifacts without finishing surface treatments (Sparacello et al. 2021).

3.4 Detection of Antimicrobial Activity: In Vitro Analysis

The activity of antimicrobial agents must be tested considering specific factors, such as reproducibility, accuracy and ease of use, as well as costs. All standardized methods used in clinical microbiology laboratories consist in *susceptibility tests* against antimicrobial agents, also detecting resistance in bacterial isolates (Jorgensen and Ferraro 2009). There are many international AST (antimicrobial susceptibility testing) standards or guidelines available from organizations and professional societies (CLSI, EUCAST, OIE).[1]

Several bioassays, such as disk/well diffusion and broth/agar diffusion, can be utilized to evaluate the in vitro efficiency of a potential antimicrobial agent (Balouiri et al. 2016; Pereira et al. 2013). The *Agar disk diffusion* method was developed in 1940, representing the routine antimicrobial susceptibility test (Balouiri et al. 2016). According to Bauer et al. (1966), this method consists in placing paper disks (6 mm in diameter) soaked with an antimicrobial agent, onto a microbial lawn growth on agar plates. The antimicrobial agent diffuses from the disks into the agar medium inhibiting microbial growth, producing an inhibition halo, whose diameter is highly related to antimicrobial activity (Fig. 3.1a). This low-cost practice offers some advantages such as simplicity and facility in testing a large number of microorganisms and antimicrobial agents (Balouiri et al. 2016; Discroll et al. 2012; Jorgensen and Ferraro 2009). This method has some variants if the

[1]CLSI Clinical and Laboratory Standard Institute; EUCAST European Committee on Antimicrobial Susceptibility Testing; OIE World Organization for Animal Health.

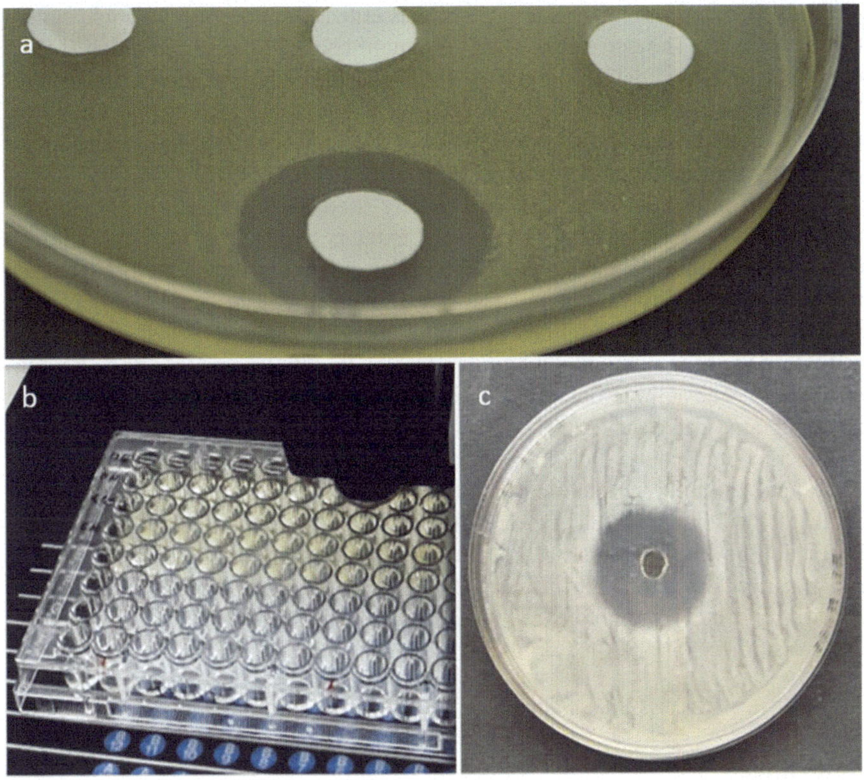

Fig. 3.1 Antimicrobial assays by (**a**) *Agar disk diffusion* method, the inhibition halo is clearly visible around the disk previously soaked with antimicrobial agent; (**b**) *Microdilution* method performed in 96-well microtiter plate; (**c**) *Well plate diffusion* method, a hole is punched into agar plate, and the halo related to fungal growth inhibition is evident around the hole (performed by aseptic method)

antimicrobial activity of natural extracts and organic acids is evaluated against bacteria and fungi. In the specific case of natural extracts or essential oils (volatile compounds), the *Aromatogram* method can be used for the evaluation of antimicrobial activity (Calvo et al. 2001).

The broth and agar dilution methods are used to determine the MIC concentration (expressed in μg/mL or mg/L) of the antimicrobial compound able to inhibit any visible microbial growth. In the case of broth dilution, test microorganisms are inoculated into a liquid growth medium with different concentrations of an antimicrobial agent (Balouiri et al. 2016; Wiegand et al. 2008). The MIC value is measured after incubation for 16–20 h (bacteria) or 36–48 h (fungi).

The broth dilution procedure (*macrodilution*) involves dilution of an antimicrobial agent in minimum 2 mL of liquid medium (Balouiri et al. 2016; Jorgensen and Ferraro 2009). The method may be performed also using 50–100 μl (*microdilution*), generally in 96-well microtiter plates (Fig. 3.1b) (Domig et al. 2007). In both methods, standardized microbial suspension is utilized to inoculate the liquid culture (NCCLS 2000). The determination of minimum bactericidal concentration (MBC) or minimum fungicidal concentration (MFC) is the most common estimation of bactericidal or fungicidal activity. The MBC is defined as the lowest concentration of an antimicrobial agent needed to kill 99.9% of the final inoculum after incubation for 24 h under a standardized set of conditions (Balouiri et al. 2016).

The *well plate diffusion* method is widely used to evaluate, in particular, the antimicrobial activity of plant or microbial extracts (Guiamet et al. 2008; Borrego et al. 2012; Gómez de Saravia et al. 2008). A hole (diameter 6–8 mm) is punched aseptically onto the surface of inoculated agar plates (Fig. 3.1c), and a volume of the antimicrobial agent, at a defined concentration, is put into the well, which then diffuses in the agar medium, inhibiting microbial growth (Calvo et al. 2001; Balouiri et al. 2016). The antimicrobial activity of plant extracts has been assayed also using aromatogram (Calvo et al. 2001).

Other methods in use include *antimicrobial gradient method* (Etest), based on the creation of an antimicrobial concentration gradient in the agar medium (Balouiri et al. 2016; Jorgensen and Ferraro 2009); *Time kill test*, used to evaluate the interaction between antimicrobial agent and microbial strain; and *ATP bioluminescence assay*, based on the measure of adenosine triphosphate (ATP) present in bacteria or fungi in order to estimate the microbial consortia (Sanmartín et al. 2016; Balouiri et al. 2016).

References

Abdel-Maksoud G, El-Amin A-R, Afifi F (2014) Insecticidal activity of Cinnamomum cassia extractions against the common Egyptian mummies' insect pest (Dermestes maculatus). Int J Conserv Sci 5(3):355–368

Acharya D, Rios JL, Rai M (2011) Naturally occurring biocides in the food industry. In: Rai M, Chikindas M (eds) Natural antimicrobials in food safety and quality. CABI E-books on agriculture and the applied life sciences from CAB international, pp 1–9. https://doi.org/10.1079/9781845937690.0001

Adamo M, Magaudda G, Nisini PT, Tronelli G (2003) Susceptibility of cellulose to attack by cellulolytic microfungi after gamma irradiation and ageing. Restaurator 24(3):145–151. https://doi.org/10.1515/REST.2003.145

Afifi HAM (2012) Comparative efficacy of some plant extracts against fungal deterioration of stucco ornaments in the Mihrab of Mostafa Pasha Ribate, Cairo. Egypt Am J Biochem Mol Biol 2(1):40–47. https://doi.org/10.3923/ajbmb.2011

Aflori M, Simionescu B, Bordianu I-E, Sacarescu L, Varganici C-D, Droftei F, Nicolescu A, Olaru M (2013) Silsesquioxane-based hybrid nanocomposites with methacrylate units containing titania and/or silver nanoparticles as antibacterial/antifungal coatings for monumental stones. Mat Sci Eng B 178(19):1339–1346. https://doi.org/10.1016/j.mseb.2013.04.004

Al-Hussaini R, Mahasneh AM (2009) Microbial growth and quorum sensing antagonist activities of herbal plants extracts. Molecules 14(9):3425–3435. https://doi.org/10.3390/molecules14093425

Alonso D, Khalil Z, Satkunanthan N, Livett BG (2003) Drugs from the sea: conotoxins as drug leads for neuropathic pain and other neurological conditions. Mini Rev Med Chem 3(7):785–787. https://doi.org/10.2174/1389557033487746

Anand U, Jacobo-Herrera N, Altemimi A, Lakhssassi N (2019) A comprehensive review on medicinal plants as antimicrobial therapeutics: potential avenues of biocompatible drug discovery. Meta 9(11):258. https://doi.org/10.3390/metabo9110258

Arias-Estévez M, López-Periago E, Martínez-Carballo E, Simal-Gándara J, Juan-Mejuto JC, García-Río L (2008) The mobility and degradation of pesticides in soils and the pollution of groundwater resources. Agric Ecosyst Environ 123(4):247–260. https://doi.org/10.1016/j.agee.2007.07.011

Ascaso C, Wierzchos J, Souza-Egipsy V, de los Rios A, Delgado Rodrigues J (2002) In situ evaluation of the biodeteriorating action of microorganisms and the effects of biocides on carbonate rock of the Jeronimos Monastery (Lisbon). Int Biodeterior Biodegradation 49(1): 1–12. https://doi.org/10.1016/S0964-8305(01)00097-X

Bacci L, Lima JKA, Araùjo APA, Blank AF, Silva IMA, Santos AA, Santos ACC, Alves PB, Picanço MC (2015) Toxicity, behavior impairment, and repellence of essential oils from pepper-rosmarin and patchouli to termites. Entomol Exp Appl 156(1):66–76. https://doi.org/10.1111/eea.12317

Balouiri M, Sadiki M, Ibnsouda SK (2016) Methods for *in vitro* evaluating antimicrobial activity: a review. J Pharm Anal 6(2):71–79. https://doi.org/10.1016/j.jpha.2015.11.005

Barresi G, Carlo E, Trapani MR, Parisi MG, Chillè C, Mulè MF, Cammarata M, Palla F (2015) Marine organisms as source of bioactive molecules applied in restoration projects. Herit Sci 3(3):17–20

Bastian F, Alabouvette C (2009) Lights and shadows on the conservation of a rock art cave: the case of Lascaux cave. Int J Speleol 38(1):55–60. https://doi.org/10.5038/1827-806X.38.1.6

Bauer AW, Kirby WM, Sherris JC, Turck M (1966) Antibiotic susceptibility by standardized single disk method. Am J Clin Pathol 45(4):493–496

Bellissima F, Bonini M, Giorgi R, Baglioni P, Barresi G, Mastromei G, Perito B (2014) Antibacterial activity of silver nanoparticles grafted on stone surface. Environ Sci Pollut Res 21(23):13278–13286. https://doi.org/10.1007/s11356-013-2215-7

Bergmann W, Feeney RJ (1950) The isolation of a new thymine pentoside from sponges. J Am Chem Soc 72(6):2809–2810. https://doi.org/10.1021/ja01162a543

Bergmann W, Feeney RJ (1951) Contributions to the study of marine products. XXXII. The nucleosides of sponges. I. J Org Chem 16(6):981–987. https://doi.org/10.1021/jo01146a023

Blanchoud H, Farrugia F, Mouchel JM (2004) Pesticide uses and transfers in urbanised catchments. Chemosphere 55(6):905–913

Blockley A, Elliott DR, Roberts AP, Sweet M (2017) Symbiotic microbes from marine invertebrates: driving a new era of natural product drug discovery. Diversity 9(4):49. https://doi.org/10.3390/d9040049

Blunt JW, Copp BR, Munro MHG, Northcote PT, Prinsep MR (2006) Marine natural products. Nat Prod Rep 23:26–78

Borderie F, Tête N, Cailhol D, Alaoui-Sehmer L, Bousta F, Rieffel D, Aleya L, Alaoui-Sossé B (2014) Factors driving epilithic algal colonization in show caves and new insights into combating biofilm development with UV-C treatments. Sci Total Environ 484:43–52. https://doi.org/10.1016/j.scitotenv.2014.03.043

Borrego S, Valdés O, Vivar I, Lavin P, Guiamet P, Battistoni P, Gómez de Saravia S, Borges P (2012) Essential oils of plants as biocides against microorganisms isolated from Cuban and Argentine documentary heritage. ISRN Microbiol 2012:1–7. https://doi.org/10.5402/2012/826786

Brogden KA (2005) Antimicrobial peptides: pore formers or metabolic inhibitors in bacteria? Nat Rev Microbiol 3(3):238–250. https://doi.org/10.1038/nrmicro1098

Calvo MA, Arosemena EL, Shiva C, Adelantado C (2001) Antimicrobial activity of plant natural extracts and essential oils. In: Méndez-Villas A (ed) Science against microbial pathogens: communicating current research and technological advances. Formatex Research Center, pp 1179–1185

Caneva G, Nugari MP, Salvadori O (1994) La biologia nel restauro. Nardini Editore, Firenze

Caneva G, Nugari MP, Salvadori O (2008) Plant biology for cultural heritage: biodeterioration and conservation. Getty Publications, Los Angeles

Cappitelli F, Abbruscato P, Foladori P, Zanardini E, Ranalli G, Principi P, Villa F, Polo A, Sorlini C (2009) Detection and elimination of cyanobacteria from frescoes: the case of the St. Brizio Chapel (Orvieto Cathedral, Italy). Microb Ecol 57(4):633–639. https://doi.org/10.1007/s00248-008-9441-4

Cappitelli F, Villa F, Sorlini C (2011) New environmentally friendly approaches against biodeterioration of outdoor cultural heritage. In: Charola AE, McNamara C, Koestler RJ (eds) Biocolonization of stone: control and preventive methods. Proceedings from the MCI workshop Series. Smithsonian contribution to museum conservation 2. Smithsonian Institution Scholarly Press, Washington DC, pp 51–58

Carrillo-González R, Martínez-Gómez MA, González-Chávez MD, Mendoza Hernández JC (2016) Inhibition of microorganisms involved in deterioration of an archaeological site by silver nanoparticles produced by a green synthesis method. Sci Total Environ S0048–9697(16): 30320–30325. https://doi.org/10.1016/j.scitotenv.2016.02.110

Carté BK (1996) Biomedical potential of marine natural products. Bioscience 46(4):271–287. https://doi.org/10.2307/1312834

Casiglia S, Bruno M, Senatore F (2014) Activity against microorganisms affecting cellulosic objects of the volatile constituents of *Leonotis nepetaefolia* from Nicaragua. Nat Prod Commun 9(11):1637–1639. https://doi.org/10.1177/1934578X1400901127

Casiglia S, Bruno M, Scandolera E, Senatore F, Senatore F (2019) Influence of harvesting time on composition of essential oil of *Thymus capitatus* (L.) Hoffmanns. & Link. growing wild in northern Sicily and its activity on microorganisms affecting historical art crafts. Arab J Chem 12(8):2704–2712. https://doi.org/10.1016/j.arabjc.2015.05.017

Chapman JS (2003) Biocide resistance mechanism. Int Biodet Biodeg 51(2):133–138. https://doi.org/10.1016/S0964-8305(02)00097-5

Chun W, DeJun Y, ShiLin H, Qun T (2000) Determination of toxicity of plant essential oils to museum insect pests. J Southwest Agric Univ 22(6):494–495

Ćirković I, Jovalekić M, Jegorović B (2012) *In vitro* antibacterial activity of garlic and synergism between garlic and antibacterial drugs. Arch Biol Sci 64(4):1369–1375. https://doi.org/10.2298/ABS1204369C

D'Agostino G, Giambra B, Palla F, Bruno M, Badalamenti N (2021) The application of the essential oils of *Thymus vulgaris* L. and *Crithmum maritimum* L. as biocidal on two Tholu Bommalu Indian leather puppets. Plan Theory 10:1508. https://doi.org/10.3390/plants10081508

Damalas CA, Eleftherohorinos I (2011) Pesticide exposure, safety issues, and risk assessment indicators. Int J Environ Res Public Health 8(5):1402–1419. https://doi.org/10.3390/ijerph8051402

Datta D, Talapatra SN, Swarnakar S (2015) Bioactive compounds from marine invertebrates for potential medicines - an overview. Int Lett Nat Sci 34:42–61. https://doi.org/10.18052/www.scipress.com/ILNS.34.42

De la Paz J, Larionova M, Maceira MA, Borrego SF, Echevarría E (2006) Control of biodeterioration using a fraction isolated from leaves of Ricinus communis linn. Pharmacology 3:462–466

Destoumieux- Garzón D, Rosa RD, Schmitt P, Barreto C, Vidal-Dupiol J, Mitta G, Gueguen Y, Bachère E (2016) Antimicrobial peptides in marine invertebrate health and disease. Philos Trans R Soc B 371(1695):20150300. https://doi.org/10.1098/rstb.2015.030

Discroll AJ, Bhat N, Karron RA, O'Brien KL, Murdoch DR (2012) Disk diffusion bioassay for the detection of antibiotic activity in body fluids: applications for the Pneumonia etiology research for child health project. Clin Infect Dis 54(2):S159–S164. https://doi.org/10.1093/cid/cir1061

Ditaranto N, van der Werf ID, Picca RA, Sportelli C, Gianossa LC, Bonerba E, Tantillo G, Sabbatini L (2015) Characterization and behaviour of ZnO-based nanocomposites designed for the control of biodeterioration of patrimonial stoneworks. New J Chem 39:6836–6843. https://doi.org/10.1039/C5NJ00527B

Doehne E, Price CA (2010) Stone conservation: an overview of current research. The Getty Conservation Institute, Los Angeles

Domig KJ, Mayrhofer S, Zitz U, Mair C, Petersson A, Amtmann E, Mayer EK, Kneifel W (2007) Antibiotic susceptibility testing of *Bifidobacterium thermophilum* and *Bifidobacterium pseudolongum* strains: broth microdilution vs. agar disc diffusion assay. Int J Food Microbiol 120:191–195. https://doi.org/10.1016/j.ijfoodmicro.2007.07.064

Donia M, Hamann MT (2003) Marine natural products and their potential applications as anti-infective agents. Lancet Infect Dis 3:338–348. https://doi.org/10.1016/s1473-3099(03)00655-8

Dubey NK, Tripathi P, Singh HB, Kumar S, Kukreja AK, Dwivedi S, Singh AK (2000) Prospects of some essential oils as antifungal agents. Int J Med Arom Plant Sci 22(1):350–354

Ebadollahi A, Ziaee M, Palla F (2020) Essential oils extracted from different species of the Lamiaceae plant family as prospective bioagents against several detrimental pests. Molecules 25(7):1556

Essa AM, Khallaf MK (2014) Biological nanosilver particles for the protection of archaeological stones against microbial colonization. Int Biodeterior Biodegradation 94:31–37. https://doi.org/10.1016/j.ibiod.2014.06.015

Faulkner DJ (2002) Marine natural products. Nat Prod Rep 19(1):1–48

Feng W, Zheng X (2007) Essential oils to control *Alternaria alternata in vitro* and *in vivo*. Food Control 18(9):1126–1130. https://doi.org/10.1016/j.foodcont.2006.05.017

Fernandes P (2006) Applied microbiology and biotechnology in the conservation of cultural heritage materials. Appl Microbiol Biotechnol 73(2):291–296. https://doi.org/10.1007/s00253-006-0599-8

Fierascu I, Ion RM, Radu M, Dima SO, Bunghez IR, Avramescu SM, Fierascu RC (2014) Comparative study of antifungal effect of natural extracts and essential oils of *Ocinum basilicum* on selected artefact. Rev Roum Chim 59(3–4):207–211

Fleeger JW, Carman KR, Nisbet RM (2003) Indirect effects of contaminants in aquatic ecosystems. Sci Total Environ 317(1–3):207–233. https://doi.org/10.1016/S0048-9697(03)00141-4

Fonseca AJ, Pina F, Macedo MF, Leal N, Romanowska-Deskins A, Laiz L, Gomez-Bolea A, Saiz-Jimenez C (2010) Anatase as an alternative application for preventing biodeterioration of mortars: evaluation and comparison with other biocides. Int Biodeterior Biodegradation 64(5): 388–396. https://doi.org/10.1016/j.ibiod.2010.04.006

Fu G, Vary PS, Lin CT (2005) Anatase TiO2 nanocomposites for antimicrobial coatings. J Phys Chem B 109(18):8889–8898. https://doi.org/10.1021/jp0502196

Ghalem BR, Mohamed B (2008) Antibacterial activity of leaf essential oils of *Eucalyptus globulus* and *Eucalyptus camaldulensis*. Afr J Pharm Pharmacol 2(10):211–215

Gómez de Saravia SG, de la Paz NJ, Guiamet P, Arenas P, Borrego S (2008) Biocide activity of natural extracts against microorganisms affecting archives. Bol Latinoam Caribe Plant Med Aromat 7(1):25–29

Gómez-Ortíz N, De la Rosa-García S, González-Gómez W, Soria-Castro M, Quintana P, Oskam G, Ortega-Morales B (2013) Antifungal coatings based on Ca(OH)2 mixed with ZnO/TiO2 nanomaterials for protection of limestone monuments. ACS Appl Mater Interfaces 5(5): 1556–1565. https://doi.org/10.1021/am302783h

Gordon EM, Barrett RW, Dower WJ, Fodor SP, Gallop MA (1994) Applications of combinatorial technologies to drug discovery. 2. Combinatorial organic synthesis, library screening strategies, and future directions. J Med Chem 37(10):1385–1401. https://doi.org/10.1021/jm00036a001

Guiamet PS, Gómez de Saravia SG (2005) Laboratory studies of biocorrosion control using traditional and environmentally friendly biocides: an overview. Latin Am Appl Res 35(4): 295–300

Guiamet PS, Gomez de Saravia SG, Arenas P, Perez ML, de la Plaz J, Borrego SF (2006) Natural products isolated from plants used in biodeterioration control. Pharmacology 3:537–544. https://doi.org/10.5402/2012/826786

Guiamet PS, de la Paz NJ, Arenas PM, Gómez de Saravia SG (2008) Differential sensitivity of *Bacillus* sp. isolated from archive materials to plant extracts. Pharmacologyonline 3:649–658

Gutarowska B, Skora J, Zduniak K, Rembisz D (2012) Analysis of the sensitivity of microorganisms contaminating museums and archives to silver nanoparticles. Int Biodeterior Biodegradation 68:7–17. https://doi.org/10.1016/j.ibiod.2011.12.002

Gutarowska B, Pietrzak K, Machnowsky W, Danielewicz D, Szynkowska M, Konca P, Surma-Slusarska B (2014) Application of silver nanoparticles for disinfection of materials to protect historical objects. Curr Nanosci 10(2):277–286. https://doi.org/10.2174/15734137113096660121

Hahn S, Schneider K, Gartiser S, Heger W, Mangelsdorf I (2010) Consumer exposure to biocide – identification of relevant source and evaluation of possible health effects. Environ Health 9(7): 2–11. https://doi.org/10.1186/1476-069X-9-7

Hamayeli H, Hassanshahian M, Askari Hesni M (2019) The antibacterial and antibiofilm activity of sea anemone (*Stichodactyla haddoni*) against antibiotic-resistant bacteria and characterization of bioactive metabolites. Int Aquat Res 11:85–97. https://doi.org/10.1007/s40071-019-0221-1

Hancock REW, Scott MG (2000) The role of antimicrobial peptides in animal defenses. Proc Natl Acad Sci U S A 97(16):8856–8861. https://doi.org/10.1073/pnas.97.16.8856

Hernández AF, Gil F, Lacasagña M (2017) Toxicological interactions of pesticide mixtures: an update. Arch Toxicol 91(10):3211–3223. https://doi.org/10.1007/s00204-017-2043-5

Jimeno JMA (2002) Clinical armamentarium of marine-derived anti-cancer compounds. Anti-Cancer Drugs 13(Suppl 1):S15–S19

Jorgensen JH, Ferraro MJ (2009) Antimicrobial susceptibility testing: a review of general principles and contemporary practices. Clin Infect Dis 49(11):1749–1755. https://doi.org/10.1086/647952

Kalemba D, Kunicka A (2003) Antimicrobial and antifungal properties of essential oils. Rev Curr Med Chem 10(10):813–829. https://doi.org/10.2174/0929867033457719

Kinghorn AD (2008) Drug discovery from natural products. In: Lemke TL, Williams DA (eds) Foye's principles of medicinal chemistry, 6th edn. Wolters Kluwer/Williams and Wilkins, Philadelphia, pp 12–25

Knetsch MLW, Koole LH (2011) New strategies in the development of antimicrobial coatings: the example of increasing usage of silver and silver nanoparticles. Polymers 3(1):340–366. https://doi.org/10.3390/polym3010340

Koestler RJ, Santoro ED (1990) Biodeterioration in museums-observations. Biodeter Res 3:505–509. https://doi.org/10.1007/2F978-1-4757-945

Koestler RJ, Parreira E, Santoro ED, Noble P (1993) Visual effects of selected biocides on easel painting materials. Stud Conserv 38(4):265–273. https://doi.org/10.2307/1506370

Kumar R, Kumar AV (1999) Biodeterioration of stone in tropical environments. An overview. The Getty Conservation Institute, Los Angeles

Kumar NS, Simon N (2016) *In vitro* antibacterial activity and phytochemical analysis of *Gliricidia sepium* (L.) leaf extracts. J Pharmacogn Phytochem 5(2):131–133

Kumar A, Kumar Vemula P, Ajayan PM, John G (2008) Silver-nanoparticles-embedded antimicrobial paints based on vegetable oil. Nat Mater 7:236–241. https://doi.org/10.1038/nmat2099

Kumar A, Shukla R, Singh P, Dubey NK (2010) Chemical composition, antifungal and antiaflatoxigenic activities of *Ocimum sanctum* L. essential oil and its safety assessment as plant based antimicrobial. Food Chem Toxicol 48(2):539–543. https://doi.org/10.1016/j.fct.2009.11.028

La Russa MF, Ruffolo SA, Rovella N, Belfiore CM, Palermo AM, Guzzi MT, Crisci GM (2012) Multifunctional TiO2 coatings for cultural heritage. Prog Org Coat 74(1):186–191. https://doi.org/10.1016/j.porgcoat.2011.12.008

Lavin P, Gómez de Saravia S, Guiamet P (2016) *Scopulariopsis* sp. and *Fusarium* sp. in the documentary heritage: evaluation of their biodeterioration ability and antifungal effect of two essential oils. Microb Ecol 71(3):628–633. https://doi.org/10.1007/s00248-015-0688-2

Lei J, Sun L, Huang S, Zhu C, Li P, He J, Mackey V, Coy DH, He Q (2019) The antimicrobial peptides and their potential clinical applications. Am J Transl Res 11(7):3919–3931

Magaudda G (2004) The recovery of biodeteriorated books and archive documents through gamma radiation: some considerations on the results achieved. J Cult Herit 5(1):113–118. https://doi.org/10.1016/j.culher.2003.07.003

Mansour M, Ahmed H (2012) Occurrence of fungi on some deteriorated ancient Egyptian materials and their controlling by ecofriendly products. EJARS 2(2):91–101. https://doi.org/10.21608/EJARS.2012.7465

Marcotte S, Estel L, Leboucher S, Minchin S (2014) Occurrence of organic biocides in the air and dust at the Natural History Museum of Rouen, France. J Cult Herit 15(1):68–72. https://doi.org/10.1016/j.culher.2013.01.005

Marshall BM, McMurry LM (2005) Biocides and resistance. In: White DG, Alekshun MN, McDermott PF (eds) Frontiers in antimicrobial resistance: a tribute to Stuart B. Levy. ASM Press, Washington, DC, pp 174–190. https://doi.org/10.1128/9781555817572.ch12

Martin-Sanchez PM, Nováková A, Bastian F, Alabouvette C, Saiz-Jimenez C (2012) Use of biocides for the control of fungal outbreaks in subterranean environments: the case of the Lascaux cave in France. Environ Sci Technol 46(7):3762–3770. https://doi.org/10.1021/es2040625

Maxim D, Bucşa L, Moza MI, Chachula O (2012) Preliminary antifungal investigation of ten biocides against moulds from two different church frescoes. Ann Rom Soc Cell Biol 17(2):139

Michaelsen A, Pinzari F, Barbabietola N, Piñar G (2013) Monitoring the effects of different conservation treatments on paper- infecting fungi. Int Biodeterior Biodegradation 84(100): 333–341. https://doi.org/10.1016/j.ibiod.2012.08.005

Mohammadi M, Masoumipour F, Hassanshahian M, Jafarinasab T (2019) Study the antibacterial and antibiofilm activity of Carum copticum against antibiotic-resistant bacteria in planktonic and biofilm forms. Microb Pathog 129:99–105. https://doi.org/10.1016/j.micpath.2019.02.002

National Committee for Clinical Laboratory Standards (2000) Methods for dilution antimicrobial susceptibility tests for bacteria that grow aerobically; approved standard M7-A5. National Committee for Clinical Laboratory Standards, Wayne 20(2)

Negi A, Sarethy IP (2019) Microbial biodeterioration of cultural heritage: events, colonization, and analyses. Microb Ecol 78:1014–1029. https://doi.org/10.1007/s00248-019-01366-y

Oliva R, Salvini A, Di Giulio G, Capozzoli L, Fioravanti M, Giordano C, Perito B (2015) TiO2-Oligoaldaramide nanocomposites as efficient core-shell system for wood preservation. J Appl Polym Sci 132(23). https://doi.org/10.1002/app.42047

Orlita A (2004) Microbial biodeterioration of leather and its control: a review. Int Biodeterior Biodegradation 53(3):157–163. https://doi.org/10.1016/S0964-8305(03)00089-1

Otero- González AJ, Magalhães BS, Garcia-Villarino M, López-Abarratequi C, Sousa DA, Dias SC, Franco OL (2010) Antimicrobial peptides from marine invertebrates as a new frontiers for microbial infection control. FASEB 24(5):1320–1334. https://doi.org/10.1096/fj.09-143388

Palla F, Bruno M, Mercurio F, Tantillo A, Rotolo V (2020) Essential oils as natural biocides in conservation of cultural heritage. Molecules 25(3):730. https://doi.org/10.3390/molecules25030730

Palla F, Caruana E, Di Carlo E, Rotolo V (2021) Plant essential oils in controlling fungal colonization on wooden substrate. Borziana 2:5–13. https://doi.org/10.7320/Borz.002.005

Patra S, Praharaj PP, Panigrahi DP, Panda B, Bhol CS, Mahapatra KK, Mishra SR, Behera BP, Jena M, Sethi G, Patil S, Patra SK, Bhutia SK (2020) Bioactive compounds from marine invertebrates as potent anticancer drugs: the possible pharmacophores modulating cell death pathways. Mol Biol Rep 47(9):7209–7228. https://doi.org/10.1007/s11033-020-05709-8

Paulus W (2005) Directory of microbicides for the protection of materials: a handbook. Springer Science & Business Media

Pereira E, Santos A, Reis F, Tavares RM, Baptista P, Lino-Neto T, Almeida-Aguiar C (2013) A new effective assay to detect antimicrobial activity of filamentous fungi. Microbiol Res 168(1):1–5. https://doi.org/10.1016/j.micres.2012.06.008

Pointing SB, Jones EBG, Jones AM (1998) Decay prevention in waterlogged archaeological wood using gamma radiation. Int Biodeterior Biodegradation 42(1):17–24. https://doi.org/10.1016/S0964-8305(98)00041-9

Pomponi SA (1999) The bioprocess-technological potential of the sea. J Biotechnol 70(1–3):5–13. https://doi.org/10.1016/S0168-1656(99)00053-X

Popovici RA, Vaduva DA, Pinzaru I, Dehelean CA, Farcas CG, Coricovac D, Danciu C, Popescu I, Alexa E, Lazureanu V, Stanca HT (2019) A comparative study on the biological activity of essential oil and total hydro-alcoholic extract of Satureja hortensis L. Exp Ther Med 18:932–942

Proksch P, Edrada R, Ebel R (2002) Drugs from the seas – current status and microbiological implications. Appl Microbiol Biotechnol 59(2–3):125–134. https://doi.org/10.1007/s00253-002-1006-8

Rai A, Prabhune A, Perry CC (2010) Antibiotic mediated synthesis of gold nanoparticles with potent antimicrobial activity and their application in antimicrobial coatings. J Mater Chem 20: 6789–6798. https://doi.org/10.1039/C0JM00817F

Rakotonirainy MS, Lavédrine B (2005) Screening for antifungal activity of essential oils and related compounds to control the biocontamination in libraries and archives storage areas. Int Biodeterior Biodegradation 55(2):141–147. https://doi.org/10.1016/j.ibiod.2004.10.002

Ranalli G, Zanardini E, Pasini P, Roda A (2003) Rapid biodeteriogen and biocide diagnosis on artworks: a bioluminescent low-light imaging technique. Ann Microbiol 53(1):1–14

Rives V, Talegon JG (2006) Decay and conservation of building stones on cultural heritage monuments. Mater Sci Forum 514:1689–1694. https://doi.org/10.4028/www.scientific.net/MSF.514-516.1689

Roman C, Diaconescu R, Scripcariu L, Grigoriu A (2013) Biocides used in preservation, restoration and conservation of the paper. Eur J Sci Theol 9(4):263–271

Rotolo V, Barresi G, Di Carlo E, Giordano A, Lombardo G, Crimi E, Costa E, Bruno M, Palla F (2016) Plant extracts as green potential strategies to control the biodeterioration of cultural heritage. Int J Conserv Sci 7(special issue 2):839–846

Ruffolo SA, La Russa MF, Malagodi M, Oliviero Rossi C, Palermo AM, Crisci GM (2010) ZnO and ZnTiO3 nanopowders for antimicrobial stone coating. Appl Phys A Mater Sci Process 100(3):829–834. https://doi.org/10.1007/s00339-010-5658-4

Ruffolo SA, Macchia A, La Russa MF, Mazza L, Urzì C, De Leo F, Barberio M, Crisci GM (2013) Marine antifouling for underwater archaeological sites: TiO2 and Ag-doped TiO2. Int J Photoenergy 2013:1–6. https://doi.org/10.1155/2013/251647

Sahab Ahmed F, Sidkey Nagwa M, Abed Nermine N, Mounir A (2014) Studies on indoor air quality in the repositories of the National Library and Archives of Egypt. IJSR 3(11):2122–2128

Şahin F, Güllüce M, Daferera D, Sökmen A, Sökmen M, Polissiou M, Agar G, Özer H (2004) Biological activities of the essential oils and methanol extract of *Origanum vulgare* ssp. *vulgare* in the eastern Anatolia region of Turkey. Food Control 15(7):549–557. https://doi.org/10.1016/j.foodcont.2003.08.009

Sakr AA, Ghaly MF, Abdel-Haliem M (2012) The efficacy of specific essential oils on yeasts isolated from the royal tomb paintings at Tanis. Egypt Int J Conserv Sci 3(2):87–92

Salem MZM, Zidan YE, Mansour MMA, El Hadidi NMN (2016) Antifungal activities of two essential oils used in treatment of three commercial woods deteriorated by five common mold fungi. Int Biodeterior Biodegradation 106:88–96. https://doi.org/10.1016/j.ibiod.2015.10.010

Salomone M, Cuttitta A, Seidita G, Mazzola S, Bertuzzi F, Ricordi C, Ghersi G (2012) Characterization of collagenolytic/proteolytic marine enzymes. Chem Eng Trans 27:1–6. https://doi.org/10.3303/CET1227001

Sanmartín P, DeAraujo A, Vasanthakumar A (2016) Melding the old with the new: trends in methods used to identify, monitor, and control microorganism on cultural heritage materials. Microb Ecol 71(4):1–17. https://doi.org/10.1007/s00248-016-0770-4

Sasso S, Scrano L, Ventrella E, Bonomo MG, Crescenzi A, Salzano G, Bufo SA (2013) Natural biocides to prevent the microbial growth on cultural heritage. In: Boriani M (ed) Proceedings of the conference built heritage 2013 monitoring conservation management. Milano, pp 1035–1042

Savković ZD, Stupar MČ, Grbić LMV, Vukojević JB (2016) Comparison of anti-*aspergillus* activity of *Origanum vulgare* L. essential oil and commercial biocide based on silver ions and hydrogen peroxide. Acta Bot Croat 75(1):121–128. https://doi.org/10.1515/botcro-2016-0011

Schieweck A, Delius W, Siwinski N, Vogtenrath W, Genning C, Salthammer T (2007) In atmospheric environment, occurrence of organic and inorganic biocides in the museum environment. Proc Indoor Air 2005 – 10th Int Conf Indoor Air Qual Climate (Part I) 41(15):3266–3275

Sequeira S, Cabrita EJ, Macedo MF (2012) Antifungals on paper conservation: an overview. Int Biodeterior Biodegradation 74:67–86. https://doi.org/10.1016/j.ibiod.2012.07.011

Silva M, Rosado T, Teixeira D, Candeias A, Caldeira AT (2015) Production of green biocides for cultural heritage. Novel biotechnological solutions. Int J Conserv Sci 6:519–530

Smith VJ, Desbois AP, Dyrynda EA (2010) Conventional and unconventional antimicrobials from fish, marine invertebrates and micro-algae. Mar Drugs 8(4):1213–1262. https://doi.org/10.3390/md8041213

Soković M, Vukojević J, Marin P, Brkić D, Vajs V, Griensven LJLD (2009) Chemical composition of essential oils of *Thymus* and *Mentha* species and their antifungal activities. Molecules 14(1): 238–249. https://doi.org/10.3390/molecules14010238

Sondossi M (2009) Biocides (nonpublic health, nonagricultural antimicrobials). In: Schaechter M, Lederberg J (eds) Desk encyclopedia of microbiology. Academic, Amsterdam

Sparacello S, Gallo G, Faddetta T, Megna B, Nicotra G, Bruno B, Giambra B, Palla F (2021) *Thymus vulgaris* essential oil and hydro-alcoholic solutions to counteract wooden artwork microbial colonization. Appl Sci 11:8704. https://doi.org/10.3390/app11188704

Speranza M, Wierzchos J, de los Rios A, Perez-Ortega S, Souza-Egipsy V, Ascaso C (2012) Towards a more realistic picture of in situ biocide actions: combining physiological and microscopy techniques. Sci Total Environ 439:114–122. https://doi.org/10.1016/j.scitotenv.2012.09.040

Sterflinger K, Pinzari F (2012) The revenge of time: fungal deterioration of cultural heritage with particular reference to books, paper and parchment. Minireview Environ Microbiol 14(3): 559–566. https://doi.org/10.1111/j.1462-2920.2011.02584.x

Ströfer-Hua E (1993) Chemicals: interaction in art, humans and nature. Restaurator 14(2):57–77. https://doi.org/10.1515/rest.1993.14.2.57

Stupar M, Grbić MLJ, Džamić A, Unkovia N, Ristić M, Jelikić A, Vukojević J (2014) Antifungal activity of selected essential oils and biocide benzalkonium chloride against the fungi isolated from cultural heritage objects. S Afr J Bot 93:118–124. https://doi.org/10.1016/j.sajb.2014.03.016

Tomei F, Baccolo TP, Papaleo B, Biagi M, Signorini S, Persechino B, Rosati MV (1996) Effects of low-dose solvents on blood of art restorers. J Occup Health 38(4):190–195. https://doi.org/10.1539/joh.38.190

Trapani MR, Parisi MG, Maisano M, Mauceri A, Cammarata M (2015) Old weapons for new wars: bioactive molecules from cnidarian internal defense systems. Cent Nerv Syst Agents Med Chem. https://doi.org/10.2174/1871524915666150710120650

Tretiach M, Crisafulli P, Imai N, Kashiwadani H, Hee Moon K, Wada H, Salvadori O (2007) Efficacy of a biocide tested on selected lichens and its effects on their substrata. Int Biodeterior Biodegradation 59(1):44–54. https://doi.org/10.1016/j.ibiod.2006.06.027

Valgas C, Machado de Souza S, Smânia EFA, Smânia AJ (2007) Screening methods to determine antibacterial activity of natural products. Braz J Microbiol 38:369–380

van der Werf ID, Ditaranto N, Picca RA, Sportelli MC, Sabbatini L (2015) Development of a novel conservation treatment of stone monuments with bioactive nanocomposites. Herit Sci 3(29): 1–9. https://doi.org/10.1186/s40494-015-0060-3

Velikova T, Trepova E, Rozen T (2011) The use of biocides for the protection of library documents: before and now. In: Méndez-Vilas A (ed) Science against microbial pathogens: communicating current research and technological advances. Formatex, Badajoz, pp 152–159

Vukojević J, Grbić ML (2010) Moulds on paintings in Serbian fine art museums. Afr J Microbiol Res 4(13):1453–1456

Walentowska J, Foksowicz-Flaczyk J (2013) Thyme essential oil for antimicrobial protection of natural textiles. Int Biodeterior Biodegradation 84:407–411. https://doi.org/10.1016/j.ibiod.2012.06.028

Walsh Z, Janeček ER, Jones M, Scherman OA (2016) Natural polymers as alternative consolidants for the preservation of waterlogged archaeological wood. Stud Conserv. https://doi.org/10.1179/2047058414Y.0000000149

Wang G (2014) Human antimicrobial peptides and proteins. Pharmaceuticals (Basel) 7(5):545–594. https://doi.org/10.3390/ph7050545

Warscheid T (2000) Integrated concepts for the protection of cultural artifacts against biodeterioration. In: Ciferri O, Tiano P, Mastromei G (eds) Of microbes and art, the role of microbial communities in the deterioration and protection of cultural heritage. Kluwer Academic/Plenum Publishers, New York, pp 185–201

Warscheid T (2003) The evaluation of biodeterioration processes on cultural objects and approaches for their effective control. In: Koestler RJ (ed) Art, biology, and conservation: biodeterioration of works of art. Metropolitan Museum of Art, New York, pp 14–27

Weinheimer AJ, Spraggins RL (1969) The occurrence of two new prostaglandin derivatives (15-epi-PGA and its acetate, methyl ester) in the gorgonian *Plexaura homomalla* chemistry of coelenterates. XV Tetrahedron Lett 59:5185–5188

Wiegand I, Hilpert K, Hancock REW (2008) Agar and broth dilution methods to determine the minimal inhibitory concentration (MIC) of antimicrobial substances. Nat Protoc 3:163–175. https://doi.org/10.1038/nprot.2007.521

Williams GP, Babu S, Ravikumar S, Kathiresan K, Prathap SA, Chinnapparaj S, Marian NP, Alikhan SL (2007) Antimicrobial activity of tissue and associated bacteria from benthic sea anemone Stichodactyla haddoni against microbial pathogens. J Environ Biol 28(4):789–793

Yang VW, Clausen CA (2007) Antifungal effect of essential oils on southern yellow pine. Int Biodeterior Biodegradation 59(4):302–306. https://doi.org/10.1016/j.ibiod.2006.09.004

Young ME, Alakomi HL, Fortune L, Gorbushina AA, Krumbein WE, Maxwell I, McCullagh C, Robertson P, Saarela M, Valero J, Vendrell M (2008) Development of a biocidal treatment regime to inhibit biological growths on cultural heritage: BIODAM. Environ Geol 56(3):631–641. https://doi.org/10.1007/s00254-008-1455-1

Biocleaning

4

Manuela Martino, Annalisa Balloi, and Franco Palla

Abstract

Cleaning is one of the first and most important steps in conservative restoration intervention, as it removes the unwanted layers of dirt and deposit from the surface of an artefact. It must be done selectively, however, by adapting the cleaning operation to the different zones and removing successive layers of deposit without acting directly on the original materials of the surface. Generally, cleaning protocols are based on chemical or physical procedures with potential negative effects for restorers' health and/or for the materials constituting the artworks. As an alternative, solvent gels, rigid gels and resin soaps can be used for selective cleaning. In recent decades, biological cleaning has greatly improved as a result of research into biotechnologies and today plays an important role in the preservation and restoration of cultural assets. Nowadays, biocleaning by viable bacterial cells or hydrolytic enzymes represents a resource with great potential in the restoration of cultural heritage, minimising risks for artworks and for human health. New methodologies based on sulphate-reducing bacteria or bioactive molecules with hydrolytic activity have been applied as selective and safer cleaning methods in the removal of black crusts from stone surfaces or organic materials such as glue and/or adhesives, from paintings and other substrates.

M. Martino (✉) · F. Palla
Department of STEBICEF, Laboratory of Biology and Biotechnologies for Cultural Heritage, University of Palermo, Palermo, Italy
e-mail: franco.palla@unipa.it

A. Balloi
Politecnico di Milano, Piazza Leonardo da Vinci, Milan, Italy
e-mail: annalisa.balloi@polimi.it

F. Palla, G. Barresi (eds.), *Biotechnology and Conservation of Cultural Heritage*,
https://doi.org/10.1007/978-3-030-97585-2_4

Keywords

Stone column · Wall painting · Cleaning method · Black crust · Animal glue

4.1 The Action of Cleaning in Restoration Projects

In the modern concept of cultural heritage restoration, the approach of minimal intervention is one of the main cornerstones established in several codes of ethics for restorers and conservators, such as that in the professional guidelines by the European Confederation of Conservator Restorers' Organisations (E.C.C.O. 2003). In summary, this approach consists in limiting interventions at minimum level, avoiding unnecessary ones and focusing on the control and/or mitigation of the causes of the decay, following the principles of 'preventive conservation' (Tabasso 2004).

From this perspective, cleaning treatments should focus essentially on the *controlled* dislodging from the surface of only those materials and substances that may interact negatively with the integrity of the underlying layers (Cremonesi 2003), re-establishing a *reasonable* aesthetic condition of the artwork and achieving a state of *legibility* (Bonsanti 2003). This means, for example, avoiding frequent and invasive artwork *maquillages* for media art exhibitions. In order not to consider cultural heritage merely as 'consumer goods', Cremonesi declares that a 'Charter of Rights' for works of art should be drawn up, so that their constituent materials are respected and they are enjoyed to the full in a sustainable way: care for the integrity of the layers through preliminary diagnosis and highly selective methods for the removal of undesired substances should therefore become a must for any advanced restorer (Cremonesi 2003).

The selectiveness of specific cleaning methods means that direct consequences are low impact for the artwork, environment and restorers involved. This is evident when comparing conventional cleaning methods (e.g. mechanical methods) with innovative protocols. For example, aqueous methods, able to couple the mild chemical action of water with other components such as chelating agents, surfactants, enzymes or living bacterial cells, are particularly selective and performing. In fact, thanks to the synergic effect of proper buffers and highly selective components, it is possible to act only against target compounds avoiding any interaction with others that we wish to preserve. As can be seen later, research on innovative cleaning methods is now moving towards biotechnology by searching for new chemicals, exploiting the specific metabolic capabilities of viable microorganisms and enhancing the proprieties of enzymes, for the bioremediation of *polluted* artworks.

4.2 Removing Undesired Layers from Artefact Surfaces

According to the illustrated glossary on stone deterioration patterns by ICOMOS[1] (Vergès-Belmin 2008), a stone alteration is a modification of the material that does not necessarily imply a worsening of its characteristics from the point of view of conservation. Thus, we may distinguish harmful from harmless alterations: the former should be eliminated or reduced because they threaten the conservation of the artwork; the latter need to be preserved because eliminating them would be pointless or even dangerous for the sound surface.

Among the harmful alterations, an additional distinction can be made between those that are endogenous or exogenous. Endogenous alterations derive from a synergic interaction between the artwork constituents and the surrounding environment (made up of pollutants and weathering), while exogenous ones derive from the accumulation of deposits or applied films on the surface, which may or may not interact with artwork constituents. Examples of harmful endogenous alterations are black and salt crusts, typical of stone or mural paintings, originated by an alteration of the constitutive minerals in sulphate (i.e. gypsum) or nitrate. They are considered undesired layers because of their disruptive action on the sound surface, which leads to loss of legibility and the gradual dissolution of the artwork. Several cleaning methods have been adopted to mitigate their presence, but many of them fail, causing loss of the constitutive materials due to the fact that altered layers are usually strongly bonded to the substrate.

On the other hand, an altered layer called *patina nobile*, resulting from a natural ageing of artefact surface, contributes to the aesthetic significance of the artwork by inducing empathy in the spectator (Weil 1985) and is usually preserved. Furthermore, patina may represent a protection for the underlying layers from weathering and aggressive pollutants.

Biological deposits, residues from inaccurate past restoration and graffiti are usually harmful or undesired exogenous alterations, which have to be removed by cleaning, with some exceptions, for example, some past restoration or finishing treatments modifying positively by ageing, forming a protective and uniform layer on the surface, such as for oxalate films (Alessandrini 2005). Epilitic lichens showing low invasiveness on the substrate sometimes have aesthetical and historical value. Their removal, furthermore, could subsequently lead to serious damage exposing the bare surface to more aggressive biodeterioration agents such as black fungi or cyanobacteria (Pinna 2004).

This schematic discussion about alterations may conceal the complexity that each restorer faces in the cleaning process. Determining the nature and potential hazard of any single layer requires an accurate technical diagnosis and a careful historical and artistic interpretation of the artwork. A respectful and 'minimal' cleaning operation foresees a proper conservation project design (Mecchi and Sansonetti 2004) leading to modular and target-specific intervention on the artwork surface.

[1] International Council on Monuments and Sites.

4.2.1 Viable Bacterial Cells

It is only in the last few decades that microbiology applied to cultural heritage has been considered in the perspective of microbial contamination and the consequent degradation of artwork surfaces. The binomial correlation microorganism deterioration, deeply rooted in medical microbiology, dominated the scientific literature of the sector during the last century, promoting the development of specific tools for the precise identification of biodegradation agents and the study of their effects. However, microorganisms are not only decaying promoters. More recently, another field of study has focused its attention on 'good microbes' that are able to lead virtuous processes, useful in restoring practices. This field of study has been developed starting from the concept that specific microorganisms, by means of their metabolic properties, are essential to all biogeochemical cycling processes and their activity plays an important part in all facets of environmental and human life (Mapelli et al. 2012).

Moreover, as microbiology has always been a practical discipline, in the era of biotechnologies this assumes new importance due to the possibility of using new tools to manage microbes and their potentialities. The exploitation of the metabolic capabilities associated with microbes has been recently defined as microbial resource management (MRM) (Verstraete et al. 2007). According to this concept, microbial communities can be managed directly (e.g. by introducing into the environment-specific microorganisms with desired capabilities) or indirectly (for instance, by acting on environmental parameters for the growth of the desired microbial species) in order to induce a positive effect in their surroundings, such as, for example, in polluted soil and water, or on the surface of artistic objects. In fact, as microorganisms are able to degrade environmental contaminants from different sites, they can, in the same way, remove undesired substances from objects that we want to preserve.

Notably, artworks and monuments, especially outdoors, are subjected to the action of several physical and chemical factors that give rise to the accumulation, on their surface, of several undesired and harmful layers such as black crusts, caused by sulphation and/or nitration phenomena, deposition of organic substances and accumulation of acidic varnish or aged adhesive, caused by inappropriate restoration practices or negligence.

As an alternative for chemical and physical cleaning techniques new biological methods have been proposed over the years. Moncreiff and Hempel in 1970 (Moncrieff and Hempel 1970) were the first to raise the possibility of biological cleaning, describing the use of a 'biological pack' and the role of microorganisms in a poultice applied for the removal of black crusts and salts on stone. Later, Hempel (1978) noted the effectiveness of a clay poultice (sepiolite and attapuglite) containing glycerol and urea and proposed that microorganisms were at least partially responsible for the cleaning process. From the end of the 1980s, selected bacteria have been successfully employed as cleaning agents, leading to the development of a new and *green* method of restoration known as *biocleaning* (or *biorestoration* in a wider sense) (Ranalli et al. 2000). In principle, the method exploits the capability of specific bacteria to use undesired substances, such as oxidised compounds of sulphur or nitrogen as electron acceptors or, in the case of organic matter, such as a carbon source, inducing their gradual degradation. Similar to enzymatic cleaning (see Sect. 4.2.2), this procedure may be included among the aqueous methods, sharing with them a high level of selectivity and low impact for artworks (Ranalli and Sorlini 2003).

A careful selection of the appropriate microorganisms, with good performance in the removal of the undesired substances (e.g. nitrates, sulphates or organic matters), is one of the first steps in planning biorestoration strategies. Thus, the best approach in finding an effective biocleaning agent is to perform a proper chemical-physical characterisation of the decay and to isolate microorganisms from the most similar chemical-physical environment (Ranalli et al. 2000; Troiano et al. 2014). Of course, selected strains have to be non-pathogenic for humans or harmless for the environment and possibly non-spore forming to facilitate their dislodging after treatment. Moreover, in order to optimise their activity on the surface that we want to clean, bacteria need to be applied by using a matrix able to (1) provide them with the right environmental conditions, (2) keep them in contact with the alteration without interacting with the surface, and (3) be quick and easy to prepare, to apply and remove (Bosch-Roig and Ranalli 2014).

Over the years, several research groups have dealt with the different aspects of biocleaning, improving the method and in some cases leading to large-scale applications and industrial development of commercial products. In the present chapter, we will discuss several case studies set up for the removal of sulphates, nitrates and organic matter alterations from monuments and mural paintings.

4.2.1.1 Removal of Sulphate

Marble and stone sulphation occurs in the presence of moisture when sulphur dioxide, a major urban atmosphere pollutant, is converted to sulphuric acid, which reacts with marble and other soluble calcareous substrates to form gypsum (Böke et al. 1999). During gypsum crystallisation, airborne organic pollutants and carbonaceous particles accumulate on surfaces protected from rainfall and wash-out and are subsequently trapped in the newly formed mineral matrix to form a so-called *black crust* (Moropoulou et al. 1998; El-Metwally and Ramadan 2005). The cleaning of crusts is essential, not only for the conservation of deteriorated areas but also for preventing further erosion phenomena (Kapsalas et al. 2007).

In the traditional conservation approach, the main methods for the removal of black crust are mechanical and chemical and, more recently, laser cleaning treatments. Mechanical methods are largely used even if they can cause erosion of the sound stone; chemical treatments generally produce good results in a reasonable time but, because of their wide range of action, are not selective and can be dangerous for human health and the environment (Lazzarini and Laurenzi Tabasso 1986). The employment of laser is a more recent method that is spreading because of its high selectivity and faster application time, though there is still uncertainty concerning its real interaction with the different substances in the crust (Salimbeni et al. 2001).

An alternative cleaning technology employs sulphate-reducing bacteria. Sulphate-reducing bacteria (SRB) are able, in anaerobic environments, to dissociate gypsum into calcium and sulphates, the latter being reduced by the bacteria and the former, reacting with carbon dioxide, converted to new calcite (Gauri and Bandyopadhyay 1999). By reducing sulphates to hydrogen sulphide, these bacteria are able to obtain energy. Thanks to this anaerobic respiration, SRB are able to break the molecular structure of gypsum, which can be easily removed from the stone surface by using water and a soft brush, cotton wool or a sponge.

The first successful application of the anaerobic sulphate reducer *Desulfovibrio desulfuricans* was reported by Atlas (1988) and Gauri and Chowdhury (1988). They obtained a partial removal of the crust (assessed only visually) by completely immersing different samples of marble with black weathering crusts rich in gypsum, in a broth containing the SRB. Calcite was also found on all the treated surfaces, suggesting that this microbe has both the potential to clean crusted marble monuments and to regenerate calcite. Another study carried out at lab scale for the removal of sulfate was conducted by Heselmeyer et al. (1991). They applied *Desulfovibrio vulgaris* to gypsum crusts, with the combined result of a removal of the black crust and a conversion of gypsum to calcite.

In 1992, using *Desulfovibrio desulfuricans*, Guari and colleagues were able to remove the black crust after 84 h, from an entire old gypsum-encrusted marble statue previously consolidated (Gauri et al. 1992) and Kouzeli (1992), also working with marble, compared chemical pastes based on EDTA or ammonium bicarbonate and reported good results of the biological technique.

Even if the immersion system had obvious drawbacks (that we will discuss later), employing SRB as a cleaning agent was promising, and further research was carried out in order to optimise the method.

The first to employ SRB outside an underwater system were Ranalli et al. (1997). They tested different strains of *Desulfovibrio* in pure and mixed culture on Candoglia marble samples of Milan Cathedral using sepiolite (a clay mineral) as a delivery system, in an anaerobic condition. With 36 h of application, about 80% of the sulfates were removed. In this experiment, the cells had to colonise the sepiolite and this process takes a long time (10–14 days). Furthermore, the colonised clay gave problems in terms of dark precipitates, due to the presence of iron in trace. To avoid the risks of colour changes on the treated surface, a pre-washed sepiolite was used to optimising the protocol, removing the iron content. Another problem with the use of inorganic carriers was the reduced metabolic activity of the cells due to the low water availability during the biotreatment. To overcome these problems, organic carriers were considered using *D. vulgaris* and Hydrobiogel-97 in an in situ treatment on outdoor altered stones on the facade of Taverna Nuova in Popoli. The moisture was covered with a plastic film for 7 days trying to guarantee anoxic conditions, but the results showed only a 28% sulfate reducion (Ranalli et al. 2000).

Later, in 2006, Cappitelli and colleagues improved the methodology by using *Desulfovibrio vulgaris* subsp. *vulgaris* ATCC 29579 and Carbogel (CTS, Vicenza, Italy) as cell carrier (Cappitelli et al. 2006). This strain is able to reduce sulphate even under low oxygen tension, making surface application easier. On the other hand,

Carbogel guaranteed a faster and higher bacterial colonisation of the delivery system, higher water retention and better contact between the cells and surface reducing some other drawbacks encountered using sepiolite, as we will see later when discussing delivery systems. The obtained bacterial matrix was applied on a marble specimen fragment from Milan Cathedral, altered with a 2–3 mm thick black crust. The matrix was not applied directly on the crust. A Japanese paper layer soaked with phosphate buffer was placed between the treated surface and the matrix, in order to facilitate the removal of the delivery system at the end of the treatment. Finally, a plastic film was applied on top of the matrix to reduce oxygen diffusion and increase water retention. The treatment ceased only when removal of the black crust was visually satisfactory: three applications, each of 15 h, were needed. Two parameters were monitored to evaluate biocleaning effectiveness: ion-exchange chromatography and colour measurements. Analysis showed a 98% removal of sulphates in the crust.

The improved biological method was further tested in two subsequent studies that demonstrated the superiority of biocleaning against chemical and physical methods. In the first one (Cappitelli et al. 2007), the SRB-Carbogel system was compared to an ammonium carbonate-EDTA mixture to remove a black crust from a *lunetta* of Candoglia marble from the Milan Cathedral. Results obtained by optical microscopy, SEM-EDS and FTIR analysis showed that the biological procedure produced a more homogenous removal of the surface deposits and a good preservation of the noble patina under the crust. In the second one (Gioventù et al. 2011), the prototype *D. vulgaris* ATCC 29579 in Carbogel was tested against chemical (ammonium carbonate-EDTA + Tween20) and laser (1064 nm, Nd:YAG laser) methods on three different lithotypes on the external walls of the Florence Cathedral: green serpentine, red marlstone and Carrara white marble. Using the above-mentioned analysis and colour measurements, it was found that the chemical method led to nonhomogenous crust removal and occasional detachment of fragments and the laser technique left a thin yellow layer visible to the naked eye, and on Carrara marble in particular, it left a residual layer of gypsum. On the contrary, biological cleaning was satisfactory and showed none of the above-mentioned drawbacks.

In 2010, *D. vulgaris* was applied for the first time on two limestone sculptures situated in the courtyard of Buonconsiglio Castle in Trento (Polo et al. 2010), obtaining successful removal of the black crust after three applications. Using the same system, biocleaning has been carried out in situ for the removal of black crust from the Pietà Rondanini by Michelangelo (Cappitelli et al. 2005) (located in the Sforzesco Castle in Milan), the sculpture 'Allegoria della Morte' by Lazzerini in Florence (Gioventù and Lorenzi 2013) and some areas of the façade of S. Maria delle Grazie, in Milan, the XVth century policrome marble in Palermo (Martino et al. 2015) and altered stone surfaces in Failaka Island, Kuwait, in this case using hydrobiogel-97 as carrier (Elhagrassy and Hakeem 2018). Despite the good results, there were still some drawbacks. Thickness and chemical heterogeneity of the crust occasionally determined long treatment times and uncompleted layer removal by the action of a single selective biocleaning agent. Indeed, black crust alterations are often not only composed of sulphate deposits but also by a complex mixture of nitrate and other various compounds such as carbonate salts, apatite and proteins (Mazzoni et al. 2014). It is worth noting that in case of powdery, an incoherent stone, prolonged contact of surfaces with water, even in a gelled state, can further exacerbate the degradation (Normandin and Slaton 2005). Several strategies, depending on the nature of the crust, have been proposed to overcome these limitations. Gioventù and Lorenzi (2010) obtained faster removal by a preliminary mechanical lowering of the crust before SRB-Carbogel applications. Troiano et al. (2013) suggested for the first time an integrated approach of chemical and biological methods, coupling on a stone column affected by black crusts the effects of SRB-Carbogel with a nonionic detergent pretreatment. The coupling of the two treatments removed the black crust without affecting the original sound marble, with a 38% reduction in cleaning time. The combined method was later applied to a century-old marble statue weathered by sulphate-based crusts and grey deposits. The detergent used alone effectively removed the grey deposit but not the black crust. However, the co-treatment synergy resulted in the complete removal of the black crust layers, with the added advantages, compared to biocleaning alone, of fewer biological applications (from seven to two) and a 70% reduction in total cleaning time.

To remove black crust made of a mixture of nitrates and sulphates, Alfano et al. (2011) proposed a multilayer biosystem consisting of a Carbogel matrix enriched with *Pseudomonas pseudoalcaligenes* KF707, a nitrate-reducing strain, and *D. vulgaris* ATCC 29579. The effectiveness of this advanced system was confirmed by long-term data monitoring, after 6 years from the start of treatment.

The same multilayer strategy was attempted by Mazzoni et al. (2014) for the first time on wall paintings (in Casina Farnese, Rome) affected by a hard-to-remove inhomogeneous deposit layer. To solubilise the crust made of calcium sulphate, calcium oxalate, apatite and aged casein, researchers employed an innovative modified laponite matrix (a colloidal clay consisting of a mixture of silicates of sodium, magnesium and lithium) containing three non-spore-forming bacterial strains: *Cellulosimicrobium cellulans* (able to solubilise calcium sulphate and carbonate), *Stenotrophomonas maltophilia* (a protein degrader) and *Pseudomonas koreensis* (able to solubilise inorganic compounds and to degrade protein material). According to the authors, laponite micro-packs containing the biocleaning agents showed high effectiveness in reducing and softening the complex deposit layer, without operating limitations. It was effective in aerobic conditions and in a wide range of temperatures (from 6 to 37 °C), safe for the restorer and with no damage to the pictorial layer or the underlying noble patina. Furthermore, micro-packs showed easy application and removal on vertical surfaces and ceilings.

In view of the interesting potential impact on the market, all these studies on the bioremoval of sulfates and black crusts lead to two patents like MI2006A000776, 19_04_2006 (spin-off Micro4You, linked with Milan University) and EU ENEA-EP 3046779 'Biotechnology process for the removal of coherent deposits of organic and inorganic origin from materials and works of artistic historical interest'. In particular, the use of *D. vulgaris* ATCC 29579 in a modified polyacrylic acid as cell carrier, was the only commercial product marketed on the field of biocleaning and available to restorers (Micro4Art sulfate produced by Micro4yoU S.r.l. and distributed by Bresciani S.r.l.). Unfortunately, a reduced market diffusion and complex production system have led to a short extent on the market. Nevertheless, several biocleaning treatments have been performed on mural paintings to remove sulfates using *D. vulgaris ATCC 29579* as biocleaning agents. In 2013 Metaldi (Metaldi 2013) applied SRB with Carbogel as carrier and covered by PET, on a mural paint of S. Apollinare Church in Trento, testing the biological cleaning only on a specific pigment (azurite); later Balloi and collegues (Balloi et al. 2015) applied the same method to remove sulfate efflorescence on mural paintings from the Queen Theodolinda Chapel in Monza Cathedral. The experimentation was carried with a short contact time (approximately 2 h), in order to avoid chromatic alteration on pigments as vermillion and lapislazzuli. In 2019 Panella and others (Panella et al. 2019) compared the results of biological cleaning testing several gels and traditional cleaning agent (ammonium dicarbonate) on a mural paint of S. Pietro chapel in S. Pudenziana church in Rome. The best system to apply the Micro4Art bioformulation proved to be the one with gellan gum.

In these studies, the microorganisms have shown positive results in the removal of sulphates and the absence of interactions with the constituent materials. A limiting factor is the need for an aqueous environment and prolonged contact times which sometimes were not possible due to the fragility of the original material and its sensitivity to water. In general, it has been shown that in the case of painted surfaces, short contact times (1–2 h) and use of rigid gels such as agar or gellan are required.

4.2.1.2 Removal of Nitrate

As for sulphates, nitrate formation affects the surface of outdoor stones. Nitrogen dioxide is a by-product coming from industrial combustion installations and vehicles. In the atmosphere, it is first oxidised to N_2O_5 and then to nitric acid. When nitric acid interacts with the calcium carbonate of the stone, it produces calcium nitrate which is more soluble than the original mineral phases and causes the pulverisation of sound stones or the generation of microcracks in wall paintings (Warscheid and Braams 2000).

Nitrate salts may also arise from industrial agriculture or from the soil, climbing up the external wall of buildings, especially when bodies are buried in close proximity, as, for example, in the case of old cemeteries constructed around churches or cathedrals (Alfano et al. 2011).

Nitrate efflorescence is also frequent on indoor wall paintings as a result of biological processes, residual restoration products left on the surface because of inaccurate restoration or natural ageing processes of the painting constituents (Doehne and Price 2010).

As for sulphates, nitrates can be reduced by specific bacteria (nitrate-reducing bacteria, NRB). The first information regarding the possible removal of nitrate deposits on frescoes through biological treatments are reported by Gabrielli (1981); the author focuses on the importance to set up experimental tests to identify the optimal growth conditions (pH, cultural media and anaerobic conditions). Ranalli et al. (1996) were the first to employ nitrate-reducing bacteria (NRB) in lab trials. Real samples of Vicenza stone altered by nitrates and artificially aged samples were treated with a strain of *Pseudomonas stutzeri* delivered in sepiolite. The application lasted 30 h, with removal of 88% of nitrates.

During the BIOBRUSH (Bioremediation for Building Restoration of the Urban Stone Heritage in European States) European research programme (May et al. 2008), Matera Cathedral was a model for in situ NRB applications. Indeed, abundant soluble nitrates, causing darkening and pulverisation of the stones, affect the lower 2 m of the external walls of the church. The salts originate from the oxidation of various N-organic compounds from bodies, which were buried in the ground when the area was used as a cemetery (Alfano et al. 2011). During the first trials, Carbogel or a mortar-alginate matrix was used as a delivery system for an NRB strain of *Pseudomonas pseudoalcaligenes*. The applications showed good nitrate removal; nevertheless, a significant contribution of the carrier especially Carbogel (up to 20% of nitrate removal) was also observed. On the other hand, mortar-alginate sustained a longer NRB activity in the interface between the surface area and the matrix.

In 2013, the system based on the use of *P. stutzeri* was applied for the first time on a wall painting altered by nitrate salt efflorescence (Bosch-Roig et al. 2013a). The research was carried out for the cleaning of wall paintings in Santos Juanes church in Valencia (Spain). In this case, among tested supports, agar was chosen as the most efficient in removing salts on vertical surfaces and as the most proper and safe because it reduces the amount of water released by the application system on the painted surface. Using this technique, a reduction of 92% in nitrate efflorescence was proved by ion-exchange chromatography.

The biological removal of nitrates has not received as much attention from scientists as sulphates, possibly because the former are more soluble than sulphates and do not produce a marked and anti-aesthetic crust as the latter (Webster and May 2006). Nevertheless, as we have seen in the above-mentioned works by Alfano et al. (2011) and Mazzoni et al. (2014), their use in combination with other biocleaning agents appears promising especially for the removal of inhomogeneous and hard-to-remove black crust made of a mixture of sulphates, nitrate and other substances.

Recently, nitrate salt efflorescence has been removed using extremophilic bacteria *Halomonas campaniensis* sp. on stone samples whit agar as carrier. The use of extremophiles in biocleaning methods is innovative, these microorganisms can adapt to very low or high temperatures, extreme values of pH, and high salt concentration. In spite of that, after treatment no *H. campaniensis* cells are able to survive and so no negative effects due to metabolism of viable cells potentially remained on the treated artwork are to be expected.

4.2.1.3 Removal of Organic Matter

In addition to air pollutants and biofilms, the surfaces of man-made artistic stonework can also be altered by organic matter that has been applied, but then not completely removed, during restoration. Residuals often act as a good growth substrate for microorganisms and mycetes that destroy the surface and allow hyphae penetration (Ranalli et al. 2005).

Fourteenth-century Pisa frescoes in the Monumental Cemetery (Camposanto), painted by famous artists such as Buonamico Buffalmacco and Spinello Aretino, represent the most important example of biocleaning for the removal of aged organic matter from painted surfaces. During an air raid by the allies in 1944, some bombs fell on the cemetery damaging the frescoes that were removed from the walls using the 'tear-off' technique, in order to be restored at a later data. The 'tear-off' technique consists in the application of a gauze on the fresco surface, by using animal glue as a consolidating agent. After glue adhesion, the cloth and the fresco form one single layer, which is then detached from the wall. One of these frescoes, 'Conversion of S. Efisio and Battle', that was restored and replaced in the original site was again removed in the 1980s, using the 'tear-off' technique, since discoloured and damaged (air humidity and pollution). Twenty years later, curators attempting a second restoration found great difficulty in removing the gauze from the previous restoration, which still adhered to the paint layer on the front surface, even when treated with a mixture of the most aggressive proteolytic enzymes available on the market (Ranalli et al. 2005). The failure was ascribed to the presence of formalin responsible for the formation of insoluble compounds during the long storage (Antonioli et al. 2005). Based on previous lab trials (Ranalli et al. 2003), a suspension of *Pseudomonas stutzeri* (DSMZ 5190) was applied to the fresco embedded in hydrophilic cotton strips, which were laid over it. After 8–12 h, the bacteria were able to digest the glue, allowing the removal of the gauze (Ranalli et al. 2005). By proteomic analysis, Antonioli et al. (2005) in a lab trial showed that in the presence of glue or aged glue, *P. stutzeri* DSMZ 5190 produces caseinolytic and collagenase activity, two fundamental enzymes needed for the digestion of the main animal glue constituents. This experiment shows once again the wide versatility of bacterial metabolism. In fact, bacteria are known to produce not only constitutive but also inducible enzymes whose synthesis takes place only in the presence of a specific substrate. These enzymes can attack and degrade different types of molecules only when the bacterial cells are exposed to them, creating a regulatory effect.

However, in this case, after gauze removal from the fresco surface, to avoid direct and longer interaction between the bacterial suspension and the painted layer, a mixture of proteolytic enzymes was employed to remove the glue residues completely (Ranalli et al. 2005).

In 2013, Bosh-Roig and colleagues improved the above-mentioned method. Using agar instead of cotton as a carrier for *P. stutzeri* (DSMZ 5190), authors were able to reduce to 2 h the duration of the biocleaning on an eighteenth-century fresco, sited at the central vault of the Santos Juanes Church, in Valencia, Spain. Compared with the cotton carrier, agar reduced the amount of the required bacteria by ten times and supported their activity better. Apart from the obvious economic advantage, the shorter interaction between bacteria and the painted surface has prevented the integrated use of expensive proteolytic enzymes (Bosch-Roig et al. 2013b). Later, the same system was carried on historical oil wall paintings (Ranalli et al. 2019). The agar-gauze activated gel showed a high effectiveness to remove protein residues on Cristo che salva Pietro dalle acque—La Navicella, by G. Lanfranco 1627-28 at Vatican Museum, and l'Incarnato, by O. and G. Riminaldi, XVII century, inside the Cupola at Pisa Cathedral.

In 2012, glue removal with *P. stutzeri* was attempted on another fourteenth-century fresco (*Stories of the Holy Fathers* by Buonamico Buffalmacco) belonging to the complex of the Monumental Cemetery in Pisa (Lustrato et al. 2012). This time the aim was to remove proteinaceous material residues from past incorrect restorations. Over the years, the proteinaceous materials had caused serious alterations and, as in the previous case, had become very hard and resistant to commonly used solvents. For the first time, a fully computerised laboratory batch fermenter (20 l useful volume) was used to obtain a suitable amount of viable bacterial cell biomass for a full-scale biocleaning (about 95 m^2 of surface area). Before application, the detached fresco was laid flat, in a horizontal position and completely covered with silk tissue paper. The bacterial culture was manually applied to the paper by using a large roller, using gentle brush strokes all over the surface. Sterile cotton swabs soaked with an abundant, activated, bacterial biomass were finally applied over the fresco, adhering to the tissue paper. After preliminary testing, ex situ biorestoration treatment was performed by a single-step application lasting only 2 h. In this case, due to the short duration of the treatment, cotton was chosen as the best carrier for the bacterial suspension. After the biocleaning, the fresco was subjected to short- and medium-term monitoring to assess microbial colonisation, activity and the presence of any undesired viable *P. stutzeri* cells. The absence of any viable cells in the fresco after bio-treatment, and thus of any potential negative effects due to their metabolism, was confirmed.

In 2019 viable bacterial cells of *P. stutzeri* were used to remove polymerised animal glue on Mural paintings in Ali kadkhoda house (El Rabiemaya), in Cairo, Egypt (ElHagrassy 2019). These mural paintings were pre-consolidated and covered by medical gauze and animal glue as an adhesive after the damages of a strong earthquake in 1992. Several methods were carried out on these murals to remove the medical gauze and animal glue, but due to the long period the animal glue had polymerised and any attempts to take it off showed no positive results and caused removal of the pigments as well. The successful biocleaning process was carried in situ by direct application of viable bacterial cells and agar as a carrier on the surface of the covered mural paintings. After the application period (3 h at 35 °C) the gauze was easily removed without causing damage to pigments.

Aged animal glues can affect other materials like paper manuscripts, creating several conservation problems. Consequently, the removal of glue residues becomes an essential step in the restoration of paper artworks. The non-pathogenic, non-spore-forming and non-cellulolytic bacterial strain *Ochrobactrum* sp. TNS15E was successfully applied in agar gel for 4 h on original XVII century paper specimens. The results showed no damage to the original cellulose fibres and the absence of undesired residues. Both colorimetry and SEM analyses proved good results (Barbabietola et al. 2016).

Another application in the biocleaning field may be seen in the removal of different hard-to-remove organic substances, such as organic synthetic polymers used in restoration—such as adhesives or protective coatings—or original constituents of contemporary artworks, which cause deterioration by ageing processes.

A valid strategy for the identification of microorganisms able to remove these materials could be to isolate them directly from varnish, paint or polymer manufacturing, industrial wastewater and sludge (Chen et al. 2007; Arutchelvan et al. 2005; Saleem et al. 2008). For instance, in 2004 a yeast was isolated from a bronze statue treated with Incralac (protective coating based on acrylic resins) and its ability to deteriorate the coating was determined (McNamara et al. 2004). The potential for microbial growth to accelerate deterioration of Incralac could be considered for a biocleaning of outdoor metal monuments. In 2014, Troiano and colleagues tested for the first time the ability of five bacteria to attack a 4-year-old Paraloid B-72. Bacteria were isolated from biodeteriorated acrylic paintings or wastewater treatment plants. Despite the presence of cell clusters over the surface of the Paraloid samples inoculated with one of the strains, observations from the solubility tests, scanning calorimetry and FTIR analysis and the stereomicroscope and SEM showed that the selected bacteria were not able to attack the polymer (Troiano et al. 2014). Nevertheless, this methodology could be considered a reference for future research on the bio-removal of synthetic resin. In this field, promising is the use of *Rhodococcus* sp. ZCONT as agent in biocleaning treatments for the removal of Paraloid B 72 from marble statue (Galizzi 2015) and vynil adhesive from frescoes (Cutraro 2014).

As regards to the synthetic polymers, also the removal of graffiti needs to be mentioned. The biocleaning of graffiti's paint is hard to perform considering its variable chemical composition consisting in several types of pigments, binders, solvents, additives and other minor components. The binders, for example, can be natural, synthetic (as alkyds, acrylics, and polyvinyl acetates,) or semi-synthetic polymers (e.g., nitrocellulose). National and European projects have recently been funded to develop anti-graffiti coatings and methods of graffiti removal: GRAFFITAGE EU project (2005–2008) aimed to develop new anti-graffiti coating systems, the SICH PRIN project (2010–2011) for eco-sustainable cleaning methods with lipase, the EFFACEUR project (2011–2012), the BioRemoGraf project (2013–2015), aimed at developing a biological method based on the use of microorganisms (Sanmartín and Bosch-Roig 2019).

In this field, several studies have obtained positive results. Giacomucci and colleagues (2012) tested the ability of *Desulfovibrio desulfuricans* ATCC 13541 to degrade nitrocellulose-based binder and red graffiti spray paint (Motip—Dupli® Autocolor 5-0200) applied to laboratory glass slides using analytical techniques, like FTIR and colorimetric analyses, to monitor the process. Sanmartín et al. (2015) reported the successful use of different microbial strains (belonging to the genera *Arthrobacter, Bacillus, Gordonia, Microbacterium, Pantoea* and *Pseudomonas* and one fungal strain belonging to the genus *Alternaria*) to degrade black graffiti spray paint. Later, other species of aerobic, mesophilic and culturable bacteria were tested in the removal of silver and black graffiti spray paints (Catto et al. 2020). Another intersting study was conducted on granite and concrete samples coated with silver and black graffiti spray paints, to compare the efficacy on different materials. Three species of bacteria (*Pseudomonas stutzeri* 5190, *Aerobacter aerogenes* 13,048, and *Comamonas* sp. 700,440), were grown with powdered graffiti and applied using two different carriers (cotton and 2% agar). Visual and microscopic examination, colour and FTIR measurements revealed remarkably successful results and important improvements have been made in the development of a biocleaning protocol applicable to the in-situ cases. The successful use of microbes in the removal of graffiti paint remains, however, a work in progress (Bosch-Roig et al. 2021).

4.2.1.4 Bioactive Molecules Extracted from Bacterial Metabolism

An innovative and interesting application of biotechnologies in the cultural heritage conservation is the use of molecules extracted from microbes with certain properties (surfactants, chelating agents, biocides, etc.).

Among these studies, it is worth mentioning the application of a biosurfactant Bio-Z extracted from *Pedobacter* sp. on lab specimens and mural paintings (Loggia Inferiore della Casina Farnese, Rome, Italy). The application with Laponite RD gel for 18 h has shown an effective surfactant action with a good reduction in surface tension and an effective removal of a thick and complex layer of dust, atmospheric particulate and casein (Grimaldi 2012; Beltrani et al. 2015).

Other molecules with interesting potential are Siderophores: secondary metabolites produced by different organisms, capable of binding iron from their surrounding environment. Siderophores present complex chemistry that allows them to form strong iron-chelating complexes (Albelda-Berenguer et al. 2019). Rapti et al. (2017) tested the siderophore deferoxamine on wood and textile samples and they compared the performance of the biological compound with the conventional chelating agent as EDTA and DTPA, in gel formulations (Klucel G, Carbopol and Agarose). Deferoxamine proved capable of removing iron corrosion products more effectively than the other chelators.

Another major line of application is the use of bioactive molecules with antimicrobial properties, used as green biocides against biological colonisation of artworks. *Bacillus* sp. are emerging as promising green alternative to common biocides, due to its capacity to produce a great range of secondary metabolites with biological activity. For example, some strains of *Bacillus subtilis* and *Bacillus amyloliquefaciens* are able to produce antifungal peptides. Bioactive peptides include antifungal peptides, antifungal lipopeptides, and antimicrobial polypeptides. Microbial lipopeptide compounds are naturally produced as defense mechanisms to stress situations (Soffritti et al. 2019).

The aim of these studies is to overcome the problem associated to the toxicity of several commercial compounds. Different natural biocompounds denominated BEVOTECH-11, BEVOTECH-14 and BEVOTECH-16, were tested against the fungal population present on the mural paintings of Casas Pintadas in Evora, Portugal, and encouraging results were obtained. These biocompounds showed large spectra of fungal inhibition, evidencing that they can be a natural alternative to commercial chemical compounds.

Marin et al. (2016) tested a detergent solutions with probiotic products (PIP), consisting in spore of *Bacillus subtilis, B. megaterium* and *B. pumilus*, able to colonise and control the proliferation of other bacterial species, obtaining good result in removing a biological patina from historical brick, showing no negative interaction with the surface (absence of decohesion, discoloration, fractures or porosity increases).

The potential use of a *Bacillus*-based biocompound to prevent the biodeterioration of artworks has also been reported in a paper analyzing the microbial population contaminating a seventeenth-century oil painting, attributed to the painter Carlo Bononi (Caselli et al. 2018). In vitro growth inhibition tests have been performed on the individual bacterial/fungal species isolated from the painting and results showed that specific probiotic *Bacillus* strains (*B. subtilis, B. pumilus, B. megaterium*) can inhibit the development of all the contaminant isolates independently of their nature (bacteria and mycetes). *Bacillus* action appears mainly based on enzymatic activity and competitive antagonisms, although the production of bacteriocins and other bioactive compounds might also be involved. Recently, a study carried out by Jurado et al. (2020) on the speleothems of Nerja Cave (Spain) showed evidence that a form of natural biological control exists in caves, providing a new method to control phototrophic biofilms. Studying the in vivo formation of plaques or spots in the biofilms where the phototrophic microorganisms disappeared, the researchers assumed the presence of predation processes operated by bacteria belonging to the genera *Bacillus* and *Lysobacter*.

According to the already reported uses of *Bacillus*-based biocompounds, potential use of such *Bacillus* strains for the decontamination of artworks surfaces and the prevention of their further colonisation is suggested, although future investigation is needed, in order to monitor the safety and the effectiveness of this approach, with short-term and long-term surveillance. In particular, it is important to ascertain the absence of any possible interaction between the biological agent and the substrate and optimise the application methods evaluating the costs in comparison with chemical compounds or other common treatments (Cappitelli et al. 2020).

In conclusion, the extreme versatility of adaptation of specific bacterial strains can lead to its potential use on a wide number of different substrates to be removed from several type of artworks. This kind of biocleaning is still an object of great interest (Soffritti et al. 2019; Gioventù et al. 2020; Ranalli and Zanardini 2021), all recent researches are focusing on confirming the effectiveness of biological systems, improving methods and time of application, selecting the most efficient delivery system.

4.2.1.5 Bacterial Cell Applications

In order to implement marketable applications of biocleaning products, two main goals have to be pursued: the first is to search for mixed populations of strains with different metabolic profiles for heterogeneous degraded layers, and the second is to make a ready-to-use product with a cheap delivery system.

As we have seen previously, researchers over time have employed different materials and methods to deliver biocleaning agents on artwork surfaces. First comparative studies between various supportants were conducted by Ranalli et al. (1997) and Cappitelli et al. (2006), carriers as sepiolite, Carbogel and Hydrobiogel-97 were compared. In general, according to a recent review by Bosch-Roig et al. (2014), a carrier should have the following characteristics:

1. Be able to retain the microorganisms and provide them with the right conditions (aerobic or anaerobic) and the water they need in order to remove the cause of decay but without any damage to the artwork itself and any undesirable changes in the colour of the surface.
2. Be applicable to all types of surfaces (horizontal, vertical, oblique, rough, smooth, etc.)
3. Be quick and easy to prepare, but also easy to apply and to eliminate at the end of the treatment, and using as far as possible only a few inexpensive materials.

Apart from the pioneering experimental applications by immersion, the principal carriers commonly employed have been sepiolite, hydrobiogel-97, Carbogel, cotton wool, arbocel agar, gellan gum and laponite. It is hard to indicate the best material to employ because it mostly depends on the single circumstances encountered by the restorer, namely, nature and position of the layer to remove, state of conservation of the artwork and, of course, the physiological requirements of the biocleaner. Thus, strongly degraded and sensitive surfaces (such as wall paintings) need short applications and low water release by the carrier. In this case, delivering bacteria together with agar on Japanese paper may be the best solution. On the other hand, when a stronger and homogenous application of the treatment or a friendly environment for anaerobic bacteria is needed, it may be better to use Carbogel. Low-cost applications on pre-consolidated surfaces may be obtained using cotton wool, with low risk to the artworks and easy-to-use applications. Finally, on vertical surfaces, good results have been achieved employing micro-packs of laponite.

A biotechnological product for non-specialised end-users should be intuitive and ready to use. Despite the increasing number of scientific papers and research groups working in the field, the technological transfer on a larger scale and the availability of products ready to use are still problems to solve. Restorers should be able to prepare it quickly even in difficult environments, such as those usually found in restoration sites. Figures 4.1, 4.2 and 4.3 reported the simple steps to prepare and apply *Micro4Art sulfate*, a biocleaning market product based on sulphate-reducing bacteria in Carboneutralgel.

4.2.2 Enzymatic Cleaning

Hydrolytic enzymes represent a very helpful tool in the biocleaning of a variety of artworks, particularly to remove dirt, adhesives and other organic residues from paintings, mural paintings and paper, wooden and stone artworks (Ranalli et al. 2005; Schwarz et al. 1999; Hamed 2012; Valentini et al. 2012; Mazzuca et al. 2014; Barbabietola et al. 2016; Germinario et al. 2017). Commercial hydrolases, such as proteases, amylases, lipases and esterases, are isolated from animal (pancreas, stomach), vegetal (seeds of oats and wheat) and microbial (bacteria or fungi) sources and utilised in biocleaning treatments in order to remove specific substrates (Palla et al. 2013; Ranalli et al. 2005). Several enzymatic treatments in cleaning have been performed on different kinds of objects with good results, representing a sustainable

Fig. 4.1 Steps to prepare Micro4Art sulfate, a biocleaning market product based on sulphate-reducing bacteria in Carboneutralgel: (**a**) transfer the freeze-dried sulphate content of one sachet into the supplied jar; (**b**) add deionised water (preferably pre-reduced water) to the freeze-dried

methodology and a safer approach to the artworks as well as a valid alternative to conventional acids and alkaline products (Bosch-Roig and Ranalli 2014; Cremonesi 2013).

Since the 1970s, enzymatic cleaning has found application in conservation treatments to remove the starch paste, animal glues or protein adhesives by amylase or protease (Wendelbo and Fosse 1970; Segal and Cooper 1977), aged acrylic coatings on painting by lipase (Bellucci et al. 1999), animal glue and protein/oily binder from paintings by mixed enzymatic solutions (Makes 1982). Enzymes, such as trypsin, amylase and protease, were mainly used for the treatment of glue stains on paper (Wendelbo 1976; Segal and Cooper 1977). During the 1990s, many restorers promoted other cleaning practices by evaluating the toxicity of the chemical products used since then (Signorini 2013). The technological biocleaning approach is also based on the combined use of viable microbial cells and hydrolytic enzymes in order to obtain the total removal of undesirable layers on artwork surfaces (Ranalli et al. 2005; Bosch-Roig et al. 2013b). Preliminary to enzymatic cleaning, characterisation of the undesired layer must be determined, for example, using chromatography techniques (Barresi et al. 2015; Cremonesi 2013; Bosch-Roig et al. 2013a, b).

Moreover, enzymatic cleaning is preferable in many cases in relationship to safety and selectively to remove organic compounds under specific pH and temperature conditions. Because commercial enzymes require a temperature ≥ 37 °C, their application can represent a limitation. Another feature to be taken into account is that the inhibitors present on the artwork, such as salts, metal ions, pigments or other molecules, may induce changes by binding to the proteinaceous structure of the enzyme (Bellucci and Cremonesi 1994).

Recently, novel purified enzymes, in particular proteases and esterases from marine invertebrates, have been isolated and assayed in bio-removal tests performed on artworks and laboratory specimens. The peculiarity of these cold-active enzymes is that they can be utilised at temperatures <30 °C, without heating enzymatic solutions or artwork surfaces, in order to successfully hydrolyse aged proteinaceous (animal glue, casein binder) or Paraloid B-72 layers (Salamone et al. 2012; Palla 2013; Palla et al. 2016; Barresi et al. 2015; Giordano et al. 2018).

After biocleaning, the effective removal of undesirable layers and the presence of residues must be checked in order to evaluate potential chromatic changes in the surface (Pruteanu et al. 2014; Micheli et al. 2016; Hrdlickova Kuckova et al. 2014; Bosch-Roig et al. 2013a, b; Palla et al. 2016).

Fig. 4.1 (continued) bacteria until the indicated volume; (**c**) add the appropriate cell carrier to obtain the desired density; (**d**) apply a layer of Japanese paper on the surface to be treated using a moist brush (preferably moistened with deionised water); (**e**) apply the bacterial suspension on the Japanese paper; (**f**) cover the area with a plastic film; (**g**) remove the wrap (plastic film, bacterial formulate and Japanese paper); (**h**) rinse the stone delicately with a damp sponge or a brush, and remove eventual residuals with a cotton bud

Fig. 4.2 Biocleanign by Micro4Art sulfate (from left) before processing; application with Japanese paper and gelled solution (Carbogel), covered by transparent polyethylene sheet; result after a total of 3 applications

Fig. 4.3 Biocleanign by Micro4Art sulfate (from left): before processing; application; result after the first application

References

Albelda-Berenguer M, Monachon M, Joseph E (2019) Siderophores: from natural roles to potential applications. Adv Appl Microbiol 106:193–225. https://doi.org/10.1016/bs.aambs.2018.12.001

Alessandrini G (2005) Patine su materiali lapidei. In: Le patine: genesi, significato, conservazione. In: Workshop promosso da Mauro Matteini. Acts by Piero Tiano e Carla Pardini. Kermes Quaderni, Nardini Editore, Firenze, pp 15–28

Alfano G, Lustrato G, Belli C, Zanardini E, Cappitelli F, Mello E, Sorlini C, Ranalli G (2011) The bioremoval of nitrate and sulfate alterations on artistic stonework: the case-study of Matera cathedral after six years from the treatment. Int Biodeterior Biodegradation 65(7):1004–1011. https://doi.org/10.1016/j.ibiod.2011.07.010

Antonioli P, Zapparoli G, Abbruscato P, Sorlini C, Ranalli G, Righetti PG (2005) Art-loving bugs: the resurrection of Spinello Aretino from Pisa's cemetery. Proteomics 5(9):2453–2459. https://doi.org/10.1002/pmic.200401182

Arutchelvan V, Kanagasabai V, Nagarajan S, Muralikrishnan V (2005) Isolation and identification of novel high strength phenol degrading bacterial strains from phenol-formaldehyde resin manufacturing industrial wastewater. J Hazard Mater 128(1–3):236–242. https://doi.org/10.1016/j.jhazmat.2005.04.043

Atlas RM, Chowdhury AN, Gauri KL (1988) Microbial calcification of gypsum-rock and sulfated marble. Stud Conserv 33(3):149–153. https://doi.org/10.2307/1506207

Balloi A, Lombardi E, Troiano F, Polo A, Cappitelli F, Gulotta D, Toniolo L, Lucchini A et al (2015) Sulfate reducing bacteria as bio-cleaning agents: development of new methodologies and study cases. Conserv Sci Cult Heritage 15:109–119

Barbabietola N, Tasso F, Alisi C, Marconi P, Perito B, Pasquariello G, Sprocati AR (2016) A safe microbe-based procedure for a gentle removal of aged animal glues from ancient paper. Int Biodeterior Biodegradation 109:53–60. https://doi.org/10.1016/j.ibiod.2015.12.019

Barresi G, Di Carlo E, Trapani MR, Parisi M, Chillè C, Mulè M, Cammarata M, Palla F (2015) Marine organisms as source of bioactive molecules applied in restoration projects. Herit Sci 3(3):17–20. https://doi.org/10.1186/s40494-015-0046-1

Bellucci R, Cremonesi P (1994) L'uso degli enzimi nella conservazione e nel restauro dei dipinti. The use of enzymes in the conservation and restoration of paintings. Kermes arte e tecnica del restauro 21:45–64

Bellucci R, Cremonesi P, Pignagnoli G (1999) A note on the use of enzymes in conservation: the removal of aged acrylic resin coatings with lipase. Stud Conserv 44(4):278–281. https://doi.org/10.1179/sic.1999.44.4.278

Beltrani T, Chiavarini S, Cicero DO, Grimaldi M, Ruggeri C, Tamburini E, Cremisini C (2015) Chemical characterization and surface properties of a new bi-emulsifier produced by *Pedobacter* sp. strain MCC-Z. Int J Biol Macromol 72:1090–1096

Böke H, Gokturk EH, Caner-Saltik EN, Demirci S (1999) Effect of airborne particle on SO_2-calcite reaction. Appl Surf Sci 140(1–2):70–82. https://doi.org/10.1016/S0169-4332(98)00468-1

Bonsanti G (2003) Storia ed etica della pulitura. In: del Convegno A, Tradizionali M (eds) Innovativi nella Pulitura dei Dipinti e delle Opere Policrome Mobili di Piazzola sul Brenta, 25–26 Ottobre 2002. Il Prato, Padova, pp 14–15

Bosch-Roig P, Ranalli G (2014) The safety of biocleaning technologies for cultural heritage. Front Microbiol 5:155. https://doi.org/10.3389/fmicb.2014.00155

Bosch-Roig P, Regidor-Ros JL, Montes-Estellés R (2013a) Biocleaning of nitrate alterations on wall paintings by *Pseudomonas stutzeri*. Int Biodeterior Biodegradation 84:266–274. https://doi.org/10.1016/j.ibiod.2012.09.009

Bosch-Roig P, Regidor-Ros JL, Soriano-Sancho P, Montes-Estellés RM (2013b) Biocleaning of animal glue on wall paintings by *Pseudomonas stutzeri*. Chem Today 31(1):50–53

Bosch-Roig P, Lustrato G, Zanardini E, Ranalli G (2014) Biocleaning of cultural heritage stone surfaces and frescoes: which delivery system can be the most appropriate? Ann Microbiol 65(3): 1227–1241. https://doi.org/10.1007/s13213-014-0938-4

Bosch-Roig P, Pozo-Antonio JS, Sanmartín P (2021) Identification of the best-performing novel microbial strains from naturally-aged graffiti for biocleaning research. Int Biodeterior Biodegradation 159. https://doi.org/10.1016/j.ibiod.2021.105206

Cappitelli F, Zanardini E, Toniolo L, Abbruscato P, Ranalli G, Sorlini C (2005) Bioconservation of the marble base of the Pietà Rondanini by Michelangelo Buonarroti. GRA 7:06675

Cappitelli F, Zanardini E, Ranalli G, Mello E, Daffonchio D, Sorlini C (2006) Improved methodology for bioremoval of black crusts on historical stone artworks by use of sulfate-reducing bacteria. Appl Environ Microbiol 72(5):3733–3737. https://doi.org/10.1128/AEM.72.5.3733-3737.2006

Cappitelli F, Toniolo L, Sansonetti A, Gulotta D, Ranalli G, Zanardini E, Sorlini C (2007) Advantages of using microbial technology over traditional chemical technology in the removal of black crusts from stone surfaces of historical monuments. Appl Environ Microbiol 73(17): 5671–5675. https://doi.org/10.1128/AEM.00394-07

Cappitelli F, Cattò C, Villa F (2020) The control of cultural heritage microbial deterioration. Microorganisms 8:1542. https://doi.org/10.3390/microorganisms8101542

Caselli E, Pancaldi S, Baldisserotto C, Petrucci F, Impallaria A, Volpe L, D'Accolti M, Soffritti I et al (2018) Characterization of biodegradation in a 17th century easel painting and potential for a biological approach. PLoS One 13:e0207630

Cattò C, Sanmartín P, Gulotta D, Troiano F, Cappitelli F (2020) Bioremediation of graffiti using novel commercial bacterial strains. Sci Total Env. https://doi.org/10.1016/j.scitotenv.2020.144075

Chen J, Zhang Y, Du GC, Hua ZZ, Zhu Y (2007) Biodegradation of polyvinyl alcohol by a mixed microbial culture. Enzyme Microbiol Tech 40(7):1686–1691. https://doi.org/10.1016/j.enzmictec.2006.09.010

Cremonesi P (2003) Un approccio alla pulitura. In: del Convegno A, Tradizionali M (eds) Innovativi nella Pulitura dei Dipinti e delle Opere Policrome Mobili, Piazzola sul Brenta, 25–26 Ottobre 2002. Il Prato, Padova, pp 59–63

Cremonesi P (2013) Rigid gels and enzyme cleaning, in new insights into the cleaning paintings. In: Mecklenburg MF, Charola AE, Koestler RJ (eds) proceedings of cleaning 2010 international conference Universidad Politécnica de Valencia and Museum Conservation Institute, vol 3, Smithsonian contributions to museum conservation. Smithsonian Institution Scholarly Press, Washington, DC, pp 179–183

Cutraro (2014) Ricerca di una via biologica per la rimozione di colle viniliche da affreschi: il caso studio degli affreschi delle cappelle di Saint Jean e Saint Martial- Palazzo dei papi, Avignone Master's Degree Dissertation in Science and Technology for the Conservation of Cultural Heritage, Sapienza Rome University

Doehne E, Price A (2010) Stone conservation: an overview of current research. Technical report. Getty Conservation Institute, Los Angeles, p 158

E.C.C.O. professional guidelines: code of ethics (2003). http://www.ecco-eu.org/about-e.c.c.o./professional-guidelines.html

ElHagrassy AF (2019) Bio-restoration of mural paintings using viable cells of *Pseudomonas stutzeri* and characterization of these murals. Int J Archaeol 7(1):8–16. https://doi.org/10.11648/j.ija.20190701.12

ElHagrassy AF, Hakeem A (2018) Comparative study of biological cleaning and laser techniques for conservation of weathered stone in Failaka Island, Kuwait. Sci Cult 4:43–50

El-Metwally AA, Ramadan AB (2005) The role of air pollutants and sewage waste in acceleration of degradation of the Islamic cultural heritage of Cairo. In: Comparative risk assessment and environmental decision making, vol 38, part 3. Springer, Dordrecht, pp 363–370

Gabrielli N (1981) Proposte di ricerca. Bollettino dei Monumenti Musei e Gallerie Pontificie, pp 123–124

Galizzi V (2015) Studio e riassemblaggio della scultura lapidea di Bacco del Museo di Antichità di Torino: ricerca di un equilibrio tra fruizione e conservazione, Master's Degree Dissertation in Conservation and Restoration of Cultural Heritage, Torino University

Gauri KL, Bandyopadhyay JK (1999) Carbonate stone, chemical behaviour, durability, and conservation. Wiley, New York

Gauri KL, Chowdhury AN (1988) Experimental studies on conversion of gypsum to calcite by microbes. In: Ciabach J (ed) Proceedings of 6th International Congress on Deterioration and Conservation of Stone. Nicholas Copernicus University, pp 545–550

Gauri KL, Parks L, Jaynes J, Atlas R (1992) Removal of sulfated-crusts from marble using sulphate-reducing bacteria. In: Webster RGM (ed) Proceedings of the International conference on stone cleaning and the nature, soiling and decay mechanisms of stone, Donheadd, Edinburgh, pp 160–165

Germinario G, van der Werf ID, Palazzo G, Ros JLR, Montes-Estelles RM, Sabbatini L (2017) Bioremoval of marker pen inks by exploiting lipase hydrolysis. Prog Org Coat 110:162–171

Giacomucci L, Toja F, Sanmartín P, Toniolo L, Prieto B, Villa F, Cappitelli F (2012) Degradation of nitrocellulose-based paint by *D. desulfuricans* ATCC 13541. Biodegradation 23:705–716

Giordano A, Rotolo V, Di Carlo E, Palla F (2018) Novel esterases for cultural heritage. X AIAr National Congress, Torino, Italy, February 14-17

Gioventù E, Lorenzi PF (2010) Valutazione dell'efficacia nell'utilizzo di microrganismi per la biorimozione delle croste nere dai materiali lapidei ed approfondimento delle metodologie applicative. OPD Restauro 22:127–138

Gioventù E, Lorenzi P (2013) Bio-removal of black crust from marble surface: comparison with traditional methodologies and application on a sculpture from the Florence's English cemetery. Procedia Chem 8:123–129. https://doi.org/10.1016/j.proche.2013.03.017

Gioventù E, Lorenzi PF, Villa F, Sorlini C, Rizzi M, Cagnini A, Griffo A, Cappitelli F (2011) Comparing the bioremoval of black crusts on colored artistic lithotypes of the Cathedral of Florence with chemical and laser treatment. Int Biodeterior Biodegradation 65(6):832–839. https://doi.org/10.1016/j.ibiod.2011.06.002

Gioventù E, Ranalli G, Vittorini Orgeas E (2020) Il biorestauro. Batteri per la conservazione delle opere d'arte. Biopulitura e bioconsolidamento, Nardini Editore, Firenze. ISBN: 978-88404-0136-2

Grimaldi M (2012) Search for new surface active compounds of microbial origin in view of the development of biorestoration techniques, PhD in science for conservation of cultural heritage, University of Florence, AA. 2009/2012

Hamed SAM (2012) A preliminary study on using enzymes in cleaning archaeological wood. J Archaeol Sci 39(7):2515–2520. https://doi.org/10.1016/j.jas.2012.02.026

Hempel K (1978) The biological pack. In: International symposium on deterioration and protection of stone monuments: international symposium, Paris, 5-9 June 1978, sec. 7.6. Paris: Unesco, Rilem

Heselmeyer K, Fischer U, Krumbein KE, Warscheid T (1991) Application of *Desulfovibrio vulgaris* for the bioconversion of rock gypsum crusts into calcite. BIOforum 1:89

Hrdlickova Kuckova S, Crhova Krizkova M, Pereira CLC, Hynek R, Lavrova O, Busani T, Branco LC, Sandu ICA (2014) Assessment of green cleaning effectiveness on polychrome surfaces by MALDI-TOF mass spectrometry and microscopic imaging. Microsc Res Tech 77(8):574–585. https://doi.org/10.1002/jemt.22376

Jurado V, del Rosal Y, Gonzalez-Pimentel JL, Hermosin B, Saiz-Jimenez C (2020) Biological control of phototrophic biofilms in a show cave: the case of Nerja Cave. Appl Sci 10:3448

Kapsalas P, Zervakis M, Maravelaki-Kalaitzaki P (2007) Evaluation of image segmentation approaches for non-destructive detection and quantification of corrosion damage on stonework. Corros Sci 49(12):4415–4442. https://doi.org/10.1016/j.corsci.2007.03.049

Kouzeli K (1992) Black crust removal methods in use: their effects on Pentelic marble surfaces. In: 7th International Congress on Deterioration and Conservation of Stone, Lisbon, pp 1147–1156

Lazzarini L, Laurenzi Tabasso M (1986) Il restauro della pietra. CEDAM, Padova, p 320

Lustrato G, Alfano G, Andreotti A, Colombini MP, Ranalli G (2012) Fast biocleaning of mediaeval frescoes using viable bacterial cells. Int Biodeterior Biodegradation 69:51–61. https://doi.org/10.1016/j.ibiod.2011.12.010

Makes F (1982) Enzymatic consolidation of a painting: seventeenth century landscape from Skokloster palace. Stud Conser 27(Suppl 1):135–138. https://doi.org/10.1179/sic.1982.27.Supplement-1.135

Mapelli F, Marasco R, Balloi A, Rolli E, Cappitelli F, Daffonchio D, Borin S (2012) Mineral–microbe interactions: biotechnological potential of bioweathering. J Biotechnol 157(4):473–481. https://doi.org/10.1016/j.jbiotec.2011.11.013

Martino M, Schiavone S, Palla F, Pellegrino L, De Castro E, Balloi A (2015) Bioremoval of Sulphate layer from a 15th century polychrome marble artifact. Conserv Sci Cult Heritage 15(1):235–252. https://doi.org/10.6092/issn.1973-9494/6177

Marin E, Vaccaro C, Leis M (2016) Biotechnology applied to historic stonework conservation: testing the potential harmfulness of two biological biocides. Int J Conserv Sci 7(1):227–238

May E, Webster AM, Inkpen R, Zamarreño D, Kuever J, Rudolph C, Warscheid T, Sorlini C, Cappitelli F, Zanardini E, Ranalli G, Krage L, Vgenopoulos A, Katsinis D, Mello E, Malagodi M (2008) In: May E, Jones M, Mitchell J (eds) Heritage microbiology and science. Microbes, monuments and maritime materials. RSC Publishing, Cambridge

Mazzoni M, Alisi C, Tasso F, Cecchini A, Marconi P, Sprocati AR (2014) Laponite micro-packs for the selective cleaning of multiple coherent deposits on wall paintings: the case study of Casina

Farnese on the Palatine Hill (Rome-Italy). Int Biodeterior Biodegradation 94:1–11. https://doi.org/10.1016/j.ibiod.2014.06.004
Mazzuca C, Micheli L, Cervelli E, Basoli F, Cencetti C, Coviello T, Iannuccelli S, Sotgiu S et al (2014) Cleaning of paper artworks: development of an efficient gel-based material able to remove starch paste. ACS Appl Mater Interfaces 8:16519–16528
McNamara CJ, Breuker M, Helms M, Perry TD, Mitchell R (2004) Biodeterioration of Incralac used for the protection of bronze monuments. J Cult Herit 5:361–364. https://doi.org/10.1016/j.culher.2004.06.002
Mecchi AM, Sansonetti A (2004) Le operazioni di pulitura alla luce dei dettami del minimo intervento: prassi già consolidata o obiettivo da perseguire? In: Il minimo intervento nel restauro. Arkos, Nardini Editore, Firenze, pp 81–90
Metaldi S (2013) Applicazione sperimentale di batteri solfato riduttori sul pigmento azzurrite. In: Atti del convegno I batteri nel restauro, Il Prato
Micheli L, Mazzuca C, Palleschi A, Palleschi G (2016) Development of a diagnostic and cleaning tool for paper artworks: a case of study. Microchem J 126:32–41. https://doi.org/10.1016/j.microc.2015.11.052
Moncrieff A, Hempel K, (1970) Biological pack. In: Conservation of stone and wooden objects 1, 103–114, ICC
Moropoulou A, Bisbikou K, Torfs K, Van Grieken R, Zezza F, Macri F (1998) Origin and growth of weathering crusts on ancient marbles in industrial atmosphere. Atmos Environ 32(6):967–982. https://doi.org/10.1016/S1352-2310(97)00129-5
Normandin K, Slaton D (2005) Cleaning techniques in conservation practice. Donhead Publishing Ltd, Dorset, p 160
Palla F (2013) Bioactive molecules: innovative contributions of biotechnology to the restoration of cultural heritage. Conserv Sci Cult Herit 13:369–378
Palla F, Cammarata M, Trapani M, Ghersi G, Salamone M, Sebastianelli M (2013) Novel proteases from marine organisms with potential interest in restoration procedure. In: Rogerio-Caldera MA, Lazzari M, Cano E (eds) Science and technology for the conservation of cultural heritage. Taylor & Francis Group, London, pp 279–282. https://doi.org/10.1201/b15577-65
Palla F, Barresi G, Giordano A, Schiavone S, Trapani MR, Rotolo V, Parisi MG, Cammarata M (2016) Cold-active molecules for a sustainable preservation and restoration of historic-artistic manufacts. Int J Conserv Sci 7(Special Issue 1):239–246
Panella E, Giovannone C, Bartolini M, Fondi V (2019) Biotechnology in the conservation field: removal of sulphates using bacteria and bioconsolidation of paintings and stuccos. Conserv Sci Cult Heritage 19:157–175. https://doi.org/10.6092/issn.1973-9494/10348
Pinna D (2004) Il minimo intervento nel controllo del degrado biologico. In: Proceding of Il minimo intervento nel restauro. Nardini Editore, Siena, pp 69–74
Polo A, Cappitelli F, Brusetti L, Principi P, Villa F, Giacomucci L, Ranalli G, Sorlini C (2010) Feasibility of removing surface deposits on stone using biological and chemical remediation methods. Microb Ecol 60(1):1–14. https://doi.org/10.1007/s00248-009-9633-6
Pruteanu S, Vasilache V, Sandu ICA, Budu AM, Sandu I (2014) Assessment of cleaning effectiveness for new ecological systems on ancient tempera icon by complementary microscopy techniques. Microsc Res Tech 77(12):1060–1070. https://doi.org/10.1002/jemt.22437
Ranalli G, Sorlini C (2003) Application of microorganisms for the deteriorates surfaces recovery. Coalition 6:2–4
Ranalli G, Zanardini E (2021) Biocleaning on cultural heritage: new frontiers of microbial biotechnologies. J Appl Microbiol. https://doi.org/10.1111/jam.14993
Ranalli G, Chiavarini M, Guidetti V, Marsala F, Matteini M, Zanardini E, Sorlini C (1996) The use of microorganisms for the removal of nitrates and organic substances on artistic stoneworks. In: Riederer J (ed) Proceedings of the 8th international congress of deterioration and conservation of stone. Möller, Berlin, pp 1415–1420

Ranalli G, Chiavarini M, Guidetti V, Marsala F, Matteini M, Zanardini E, Sorlini C (1997) The use of microorganisms for the removal of sulfates on artistic stoneworks. Int Biodeterior Biodegradation 40(2–4):255–261. https://doi.org/10.1016/S0964-8305(97)00054-1

Ranalli G, Matteini M, Tosini I, Zanardini E, Sorlini C (2000) Bioremediation of cultural heritage: removal of sulphates, nitrates and organic substances. In: Ciferri O, Tiano P, Mastromei G (eds) Of microbes and art. The role of microbial communities in the degradation and protection of cultural heritage. Kluwer Academic/Plenum Publishers, New York, pp 231–245

Ranalli G, Belli C, Baracchini C, Caponi G, Pacini P, Zanardini E, Sorlini C (2003) Deterioration and bioremediation of frescoes: a case-study. In: Saiz-Jimenez C (ed) Molecular biology and cultural heritage. A.A. Balkema Publishers, Lisse, pp 243–246

Ranalli G, Alfano G, Belli C, Lustrato G, Colombini MP, Bonaduce I, Zanardini E, Abbruscato P, Cappitelli F, Sorlini C (2005) Biotechnology applied to cultural heritage: biorestoration of frescoes using viable bacterial cells and enzymes. J Appl Microbiol 98(1):73–83. https://doi.org/10.1111/j.1365-2672.2004.02429.x

Ranalli G, Zanardini E, Rampazzi L, Corti C, Andreotti A, Colombini MP, Bosch-Roig P, Lustrato G et al (2019) Onsite advanced biocleaning system on historical wall paintings using new agar gauze bacteria gel. J Appl Microbiol 126:1785–1796

Rapti S, Boyatzis S, Rivers S, Velios A, Pournou A (2017) Removing iron stains from wood and textile objects: assessing gelled siderophores as novel green chelators. In: Angelova V, Ormsby B, Townsend JH, Wolbers R (eds) Gels in conservation of art. Archetype Publications Ltd, pp 343–348

Salamone M, Cuttitta A, Seidita G, Mazzola S, Bertuzzi F, Ricordi C, Ghersi G (2012) Characterization of collagenolytic/proteolytic marine enzymes. Chem Eng 27:1–6. https://doi.org/10.3303/CET1227001

Saleem M, Brim H, Hussain S, Arshad M, Leigh MB (2008) Perspectives on microbial cell surface display in bioremediation. Biotechnol Adv 26(2):121–186. https://doi.org/10.1016/j.biotechadv.2007.10.002

Salimbeni R, Pini R, Siano S (2001) Achievement of optimum laser cleaning in the restoration of artworks: expected improvements by on-line optical diagnostics. Spectrochim Acta B At Spectrosc 56(6):877–885. https://doi.org/10.1016/S0584-8547(01)00197-5

Sanmartín P, Bosch-Roig P (2019) Biocleaning to remove graffiti: a real possibility? Advances towards a complete protocol of action. Coatings 9:104

Sanmartín P, DeAraujo A, Vasanthakumar A, Mitchell R (2015) Feasibility study involving the search for natural strains of microorganisms capable of degrading graffiti from heritage materials. Int Biodeterior Biodegrad 103:186–190. https://doi.org/10.1016/j.ibiod.2015.05.010

Schwarz I, Blüher A, Banik G, Thobois E, Maurer KH (1999) The development of a ready-for-use poultice for local removal of starch paste by enzymatic action. Restaurator 20(3–4):225–244. https://doi.org/10.1515/rest.1999.20.3-4.225

Segal J, Cooper D (1977) The use of enzymes to release adhesives. Paper Conservator 2(1):47–50. https://doi.org/10.1080/03094227.1977.9638498

Signorini E (2013) Surface cleaning of paintings and polychrome objects in Italy: the last 15 years, in new insights into the cleaning paintings. In: Mecklenburg MF, Charola AE, Koestler RJ (eds) Proceedings of cleaning 2010 international conference Universidad Politécnica de Valencia and Museum Conservation Institute, Vol 3, Smithsonian contributions to museum conservation. Smithsonian Institution Scholarly Press, Washington, DC, pp 17–23

Soffritti I, D'Accolti M, Lanzoni L, Volta A, Bisi M, Mazzacane S, Caselli E (2019) The potential use of microorganisms as restorative agents: an update. Sustainability 11:3853. https://doi.org/10.3390/su11143853

Tabasso ML (2004) Il criterio del minimo intervento: possibili interpretazioni e applicazioni alla conservazione dei materiali lapidei. Il minimo intervento nel restauro, atti del convegno, 18–19 giugno, Siena. Nardini, Firenze, pp 9–16

Troiano F, Gulotta D, Balloi A, Polo A, Toniolo L, Lombardi E, Daffonchio D, Sorlini C, Cappitelli F (2013) Successful combination of chemical and biological treatments for the cleaning of stone artworks. Int Biodeterior Biodegradation 85:294–304. https://doi.org/10.1016/j.ibiod.2013.08.011

Troiano F, Vicini S, Gioventù E, Lorenzi PF, Improta CM, Cappitelli F (2014) A methodology to select bacteria able to remove synthetic polymers. Polym Degrad Stab 107:321–327. https://doi.org/10.1016/j.polymdegradstab.2013.12.029

Valentini F, Diamantia A, Carbone M, Bauer EM, Palleschi G (2012) New cleaning strategies based on carbon nanomaterials applied to the deteriorated marble surfaces: a comparative study with enzyme based treatments. Appl Surf Sci 258(16):5965–5980. https://doi.org/10.1016/j.apsusc.2012.01.076

Vergès-Belmin V (2008) Illustrated glossary on stone deterioration patterns. Glossaire illustré sur les formes d'altération de la pierre. English-French ed, Monuments & Sites no. 15. ICOMOS and (ISCS) international Scientific Committee for Stone, Paris

Verstraete W, Wittebolle L, Heylen K, Vanparys B, de Vos P, van de Wiele T, Boon N (2007) Microbial resource management: the road to go for environmental biotechnology. Eng Life Sci 7(2):117–126. https://doi.org/10.1002/elsc.200620176

Warscheid T, Braams J (2000) Biodeterioration of stone: a review. Int Biodeterior Biodegradation 46(4):343–368. https://doi.org/10.1016/S0964-8305(00)00109-8

Webster A, May E (2006) Bioremediation of weathered-building stone surfaces. Trends Biotechnol 24(6):255–260. https://doi.org/10.1016/j.tibtech.2006.04.005

Weil P (1985) Patina: historical perspective on artistic intent and subsequent effects of time, nature and man. In: Naudé V (ed) Sculptural monuments in an outdoor environment. Pennsylvania Academy of the Fine Arts, Philadelphia, pp 21–27

Wendelbo O (1976) The use of proteolytic enzymes in the restoration of paper and papyrus. University Library of Bergen, Bergen

Wendelbo O, Fosse B (1970) Protein surgery: a restoring procedure applied on paper. Restaurator 1(4):245–248

Innovative and Integrated Strategies: Case Studies

5

Franco Palla, Giovanna Barresi, Matteo Cammarata, Giuseppe Cavallaro, Rosa Maria Chisesi, Enza Di Carlo, Ambra Giordano, Giovanna Lombardo, Maria Giovanna Parisi, Valentina Rotolo, Salvatore Schiavone, Giuseppe Arcoleo, and Giulia Casini

Abstract

In this chapter, case studies related to *biodeterioration*, *bioaerosol*, *biocide* and *biocleaning* are reported. The aim is highlighting the role of biology and biotechnology tools for the preventive conservation of organic and inorganic artefacts, understanding how traditional as well as innovative methods can help the conservationists to develop integrated strategies considering works of art/environment/humans as a dynamic system. Particularly, based on the experience acquired during the researches of *Laboratory of Biology and Biotechnology for Cultural Heritage* (*LaBBCH*), the authors suggest several approaches to reveal and identify biological systems able to induce biodeterioration of cultural assets, also focusing on bioaerosols in indoor environment to assess the risk for historical-artistic collections. Finally, novel bioactive molecules have been applied to perform biocleaning protocols or to control of microbial colonisation, in accordance with conservative restoration procedures and safety for both the

F. Palla (✉) · G. Barresi · R. M. Chisesi · E. Di Carlo · A. Giordano · G. Lombardo · V. Rotolo · S. Schiavone
Department STEBICEF, Laboratory of Biology and Biotechnologies for Cultural Heritage, University of Palermo, Palermo, Italy
e-mail: franco.palla@unipa.it; ambra.giordano@unipa.it; giovanna.lombardo@unipa.it

M. Cammarata · M. G. Parisi
Department DiSTeM, University of Palermo, Palermo, Italy
e-mail: matteo.cammarata@unipa.it; mariagiovanna.parisi@unipa.it

G. Cavallaro
Department Physics and Chemistry, University of Palermo, Palermo, Italy
e-mail: giuseppe.cavallaro@unipa.it

G. Arcoleo · G. Casini
Enbiotech S.r.l., Palermo, Italy
e-mail: g.arcoleo@enbiotech.eu; g.casini@enbiotech.eu

F. Palla, G. Barresi (eds.), *Biotechnology and Conservation of Cultural Heritage*,
https://doi.org/10.1007/978-3-030-97585-2_5

environment and operators, as well as nanomaterial (halloysite nanotubes) to consolidate and protect wooden archaeological finds.

Keywords

Benzalkonium chloride · Biological patina · Black crust · Nanotubes · Plant extract

5.1 Biodeterioration of Inorganic Cultural Assets

5.1.1 Fountains

The fountains and all decorative apparatus usually exposed to a water supply and located in outdoor may suffer of several biological colonisation. This is, for example, the case of the biological community revealed in the *Fountain of the Two Dragons* in Palermo (Fig. 5.1a) where the attention has been paid to all discolourations, encrustations and pigmented patinas traceable to past or actual biological systems living onto marble surfaces (Di Carlo et al. 2015). Plants, mosses, insects, rusty patinas, black crusts, dirty deposits, fissures and detachment of material have been investigated. In order to define an adequate conservative restoration procedure, the identification of biodeteriogens has been required and the characterisation performed through an integrated approach based on microscopy techniques, in vitro culture and molecular biology investigation. Particularly, pigmented areas have been sampled in order to reveal microbial consortia able to release biogenic pigments. All samplings have been performed using both non-invasive (sterile swab, nylon membrane fragment) and microinvasive methods (micro-scalpel) and supported by portable field microscope (Fig. 5.1b). Macro-systems, collected in different areas of the fountain, have been analysed by digital microscope (Dino-Lite) and stereomicroscope (Wild Heerbrugg); collected bryophytes and insects (or their fragments) have been identified by optical microscope observations (Leica, 40×).

Concerning the microbial community inhabiting the fountain, reddish and pinkish bacteria mostly belonged to *Arthrobacter* and *Paracoccus* genera have been

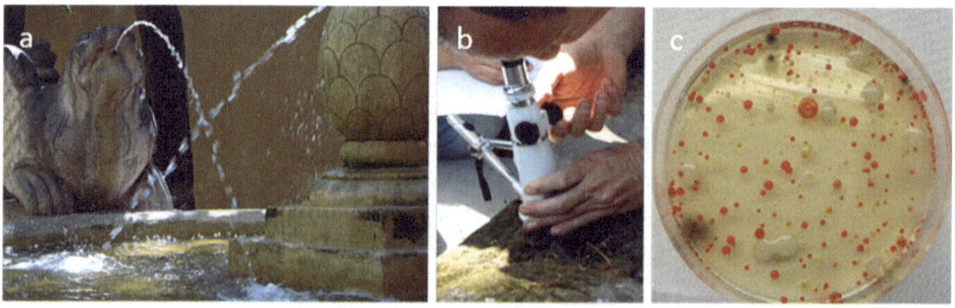

Fig. 5.1 Fountain of two dragons: (**a**) water supply in the basin; (**b**) field portable microscope; (**c**) microbial colonies developed on N-agar medium

identified (Fig. 5.1c). Molecular biology investigation has been performed through ad hoc protocols, including genomic DNA extraction by Genomic DNA Purification Kit—Fermentas (Palla 2020). The sequencing has been performed by Eurofins MWG Operon and sequencing service (Germany) and sequence analysis by BLAST platform using the nucleotide databases NCBI-NIH (USA) and EMBL (Germany) (Altschul et al. 1997). Fungal colonies belonging to *Alternaria* spp., *Fusarium* spp., *Cladosporium* spp. and *Phoma* spp. have been revealed. In particular, *Cladosporium* and *Phoma* are considered two of the most important fungal species involved in biodeterioration of urban stone buildings (Sterflinger and Prillinger 2001; Sterflinger and Piñar 2013). Further studies on algae and cyanobacteria will undertake in order to plan a successful conservative intervention.

During *Fontana Pretoria* (Palermo) restoration, a comprehensive study on colonising microbial communities has been carried out. The marble areas showed an expanding and intensifying reddish chromatic alteration. As reported in the literature (Krumbein 2003), these chromatic alterations could be related to bacterial taxa that we identify in this study by molecular investigation (Palla and Tartamella 2007). Non-invasive sampling has been carried out on marble statues by sterile cotton swabs soaked with NaCl-Tween20 solution. Bacteria as *Bacillus*, *Arthrobacter*, *Micrococcus* and *Cellulomonas* have been identified combining in vitro culture, SEM observations and molecular investigations, according to protocols carried out in our laboratory. The adequate biocide concentrations (Algophase and Benzalkonium chloride) have been determined by "inhibition bacterial growth" in in vitro assays (Palla 2009).

5.1.2 Wall Paintings

Saints Cave (XII–XIV sec., Licodia Eubea, Alia, Sicily) restoration project represented an opportunity for investigating the micro and macro-biological systems colonising this environment, characterised by high relative humidity, percolating water and a continuous air exchange with the surrounding countryside. Among macro-colonisers, *Parametria diffusa* M. et K. and *Adiantum capillus-veneris* identified in the confined environment, microbial systems able to trigger the degradation on frescoes have been revealed combining microscopy and molecular analysis. Black, green and differently pigmented areas inhabiting the cave wall surfaces have been investigated by CLSM observations (Olympus FV-300 equipped with Argon laser, $\lambda = 488$ nm, and Helium/Neon laser, $\lambda = 543$ nm) at 40× magnification, and the main colonising microorganisms (*Cyanobium*, *Aspergillus*, *Trichoderma*) have been identified by molecular investigation (Palla et al. 2010). The molecular analysis included microbial DNA extraction (Genomic DNA Purification Kit *Fermentas*), PCR (Ready-to-Go *Amersham Biosciences* reaction mixture), sequencing (Eurofins MWG Operon sequencing service) and sequence analysis performed by BLAST platform (Altschul et al. 1997).

Detection of microbial community on fresco surfaces (Fig. 5.2a) has been carried out in *Borremans* hall (Palermo *Diocesan Museum*). The goal of this investigation was to assess the microbial contamination on the painted surfaces at different heights

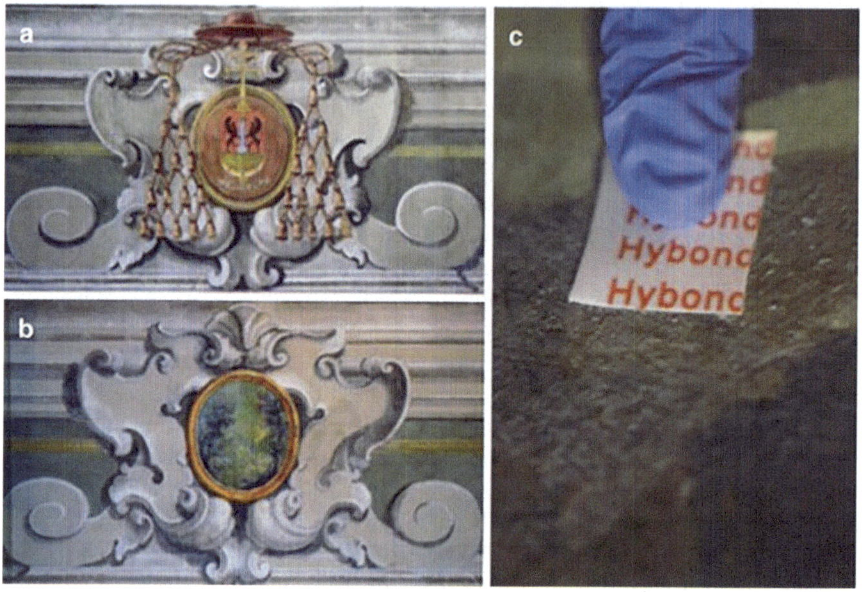

Fig. 5.2 *Borremans* frescoes: (**a**, **b**) investigated areas; (**c**) sampling by nylon membrane fragment (Hybond$^+$, Amersham)

in order to evaluate the microbial "fallout" (microbial sedimentation on a given surface) and how the microbial contamination of aerosol may influence the colonisation of artworks surfaces; for this reason, the monitoring of airborne microbial load was carried out in parallel. Non-invasive sampling (nylon membrane fragments 4 cm^2) has been utilised in order to isolate microorganisms in selected points on fresco surface (Fig. 5.2b). Morphological study of fungal colonies and DNA investigation allowed the identification of microorganisms as *Penicillium chrysogenum*, *Chaetomium globosum*, *Aspergillus versicolor*, *Cladosporium* spp., *Kocuria rosea*, *Nocardiopsis* spp., *Staphylococcus* spp., *Micrococcus luteus* and *Bacillus* spp. The higher microbial load (4 CFU/cm^2) was attributable to *Penicillium chrysogenum*.

Moreover, during the fresco samplings, the surface has been irradiated by UV light, and an elemental analysis has been performed by portable X-ray fluorescence (XRF) instrument (Drand and Laval 2009).

5.2 Biodeterioration of Organic Cultural Assets

5.2.1 Paper

Especially for those objects made of paper, the detection of fungi or bacteria is a fundamental step in order to prevent and control the paper decay. The main foresight is the occurrence of a non-destructive sampling; nitrocellulose and nylon membranes have been found to be useful for this purpose. Moreover, a constant monitoring of specific microbial strains enables the evaluation of the most appropriate conservation strategies.

In this scenario, the identification of bacteria and especially of fungi is fundamental, since they may be considered the main cause of chromatic alterations, originating spots of different colours. It has been observed that the presence of fungi can cause also structural alterations that means of fragility of paper or their partial/total destruction. It is necessary to keep in mind that a non-invasive sampling allows the isolation of bacterial and fungal community in archive, libraries and paper document repositories without damaging the cultural items. Particularly, a study of microbial deteriogens has been carried out on documents, drawings and photos. After sampling, the membranes have been put onto nutritive media (Sabouraud or N-Agar) in order to favour microbial growth (Palla 2015). The results obtained from culture-dependent method, molecular analysis and microscopic identification have confirmed the presence of fungal taxa, including *Cladosporium* spp., *Penicillium* spp. and *Aspergillus* spp. typical fungal infections in libraries, colonising documents made of paper (Sterflinger and Piñar 2013). Differently pigmented areas were sampled by nylon membranes (Fig. 5.3a) to perform culture-dependent analysis (Fig. 5.3b). The morphology of isolated fungal strains has been determined in parallel with molecular investigation, performed through the amplification of 18S-ITS1 rDNA target sequences (Palla et al. 2011). The *Penicillium* has been found as dominant genus, showing a very high microbial load; actinomycetes and bacteria (*Bacillus* and *Micrococcus*) were also found.

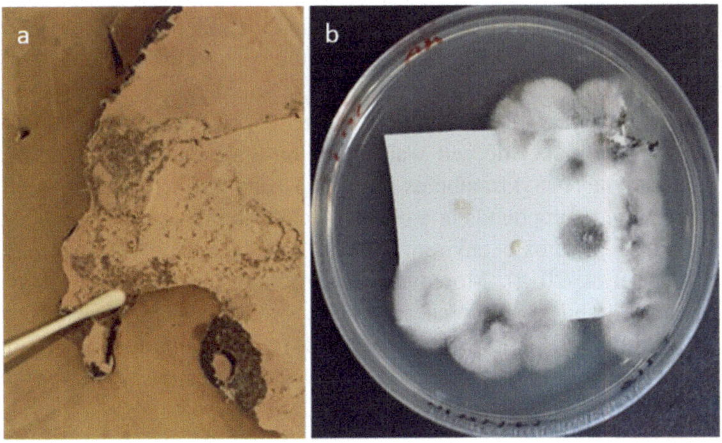

Fig. 5.3 (**a**) Paper affected by microbial decay; (**b**) wide fungal growth (*Penicillium* spp.) on nylon membrane fragment onto nutritive medium

5.2.2 Leather and Parchment

Microbial colonisation on several ancient parchment items has been investigated. Non-invasive sampling has been performed using sterile nylon membrane fragments, which were then inoculated into nutritive solid media. The surface sampling has been carried out both on previously treated by fumigation (5 years before) and not treated parchments. The study has sought to highlight the antimicrobial efficiency of fumigation treatment over time. However, the results showed that the disinfection treatment has not prevented microbial recolonisation, allowing the growth of very dangerous fungal strains, such as *Penicillium* and *Alternaria*. Furthermore, the presence of *Micrococcus* spp. and *Staphylococcus* spp. is probably attributable to wrong handling parchments by operators and users.

5.2.3 Wood

The investigation protocols developed in our studies (Palla et al. 2013, 2015a; Perez et al. 2017) have been applied to a wide range of wooden findings and artefacts, in order to shed some light on microbial degradation phenomena, indispensable for correct conservation and restoration strategies.

Archaeological waterlogged wood, part of a bronze rostrum (*Rostro di Acqualadroni*) dated from third century BC (Santamaria et al. 2009), has been analysed by an integrated approach. Particularly, cross and radial sections of wood samples have been observed by light microscopy (magnification range 10.5–40×, reflected light and 40–200× transmitted light) and identified as *Pinus* sp. SEM. The SEM investigation has been utilised to assess the changes in the wood anatomical structure and to deal the state of conservation of submerged wood, evaluating the decay traceable to microbial activity (Blanchette 2000). The SEM micrographs (Fig. 5.4) revealed a specific cell wall alteration attributable to bacterial activity and abundant pyrite (FeS_2) framboids (present as single structure or clustered). The presence of sulphur compounds in archaeological waterlogged wood indicates both long-term burial in anoxic environment and colonisation by sulphate-reducing bacteria (Palla et al. 2014). Moreover, using culture-independent methods, iron-oxidising bacteria and iron sulphate reducers, respectively, *Marinobacter* sp. and *Desulforudis audaxviator*, have been identified. Bacteria with ligninolytic or cellulolytic activity belonging to *Cellulomonas*, *Bacillus*, *Pseudomonas*, *Xanthomonas* and *Sphingomonas* genera were also identified.

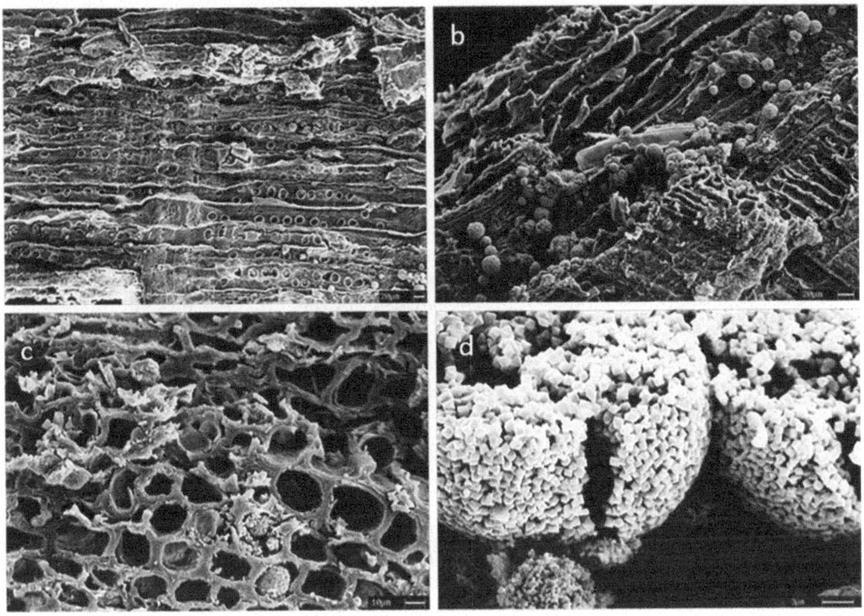

Fig. 5.4 SEM micrographs of wooden thin sections. (**a**, **b**) radial wood structure; (**c**) pyrite framboids in tracheids; (**d**) magnification of a single framboid

5.3 Bioaerosol in Confined and Semi-Confined Environments

Confined and semi-confined environments show peculiar temperature and relative humidity values, fluctuating between day and night, where microbial metabolites are excreted and gaseous and particulate pollution are present, all phenomena that play an important role in the artwork deterioration. Heating, air-conditioning or ventilating system (HVAC), building structures, outside air exchange or huge numbers of visitors generate potential risk factors. In order to establish exposure thresholds, microbial fraction of air samples must be isolated, by means of active or passive sampling.

In the last years, different indoor cultural heritage environments were investigated by the measurement of environmental and microenvironmental parameters that are essential for a correct study of the site and for its correct conservation in accordance with the typologies of material stored or exposed in.

During these studies, we emphasise the use of integrated approach to outline the microbial contamination in indoor air of Sicilian cultural sites (Table 5.1) and its role in both artefact deterioration and health effects on visitors/operators. In order to detect and characterise the microbial contamination, the sampling of surface (swab, nylon membranes) and aerosol (active, passive) has been performed to obtain complete information on microbial consortia. Particularly for hypogeal

Table 5.1 Airborne microorganisms (bacteria and fungi) identified in crypt and museum environments by active and passive method

Environment	Sampling method	Bacteria	Fungi
Crypt	AirPort MD8	*Nocardiopsis* spp. *Staphylococcus* spp.	*Alternaria* spp. *Cladosporium* spp. *Penicillium* spp.
	Sedimentation	*Arthrobacter* spp. *Bacillus subtilis* *Micrococcus luteus* *Paracoccus* spp. *Staphylococcus* spp.	*Alternaria* spp. *Chaetomium* spp. *Cladosporium* spp. *Penicillium chrysogenum* *Phoma* spp.
Museum	AirPort MD8	*Bacillus* spp. *Paracoccus* spp. *Staphylococcus* spp.	*Aspergillus* spp. *Aspergillus niger* *Cladosporium* spp. *Penicillium* spp. *P. chrysogenum* *Scopulariopsis* spp. *Trichoderma* spp.
	Sedimentation	*Arthrobacter* spp. *Bacillus* spp. *Micrococcus* spp. *Nocardiopsis* spp. *Paracoccus* spp. *Pseudomonas* spp. *Staphylococcus* spp.	*Aspergillus* spp. *Cladosporium* spp. *Penicillium* spp.

environments, their structural features mainly attributable to the lithic substrates and the different air exchanges with the outside may enhance the growth of several microbial taxa. When the environment is characterised by stable environmental parameters and reduced number of visitors, few amount of microbial taxa may be isolated (Di Carlo et al. 2016).

In museums and archives, bioaerosols generally have a different composition with respect to hypogeal environments. Usually, in archives/libraries and museums/galleries, different artefacts very often composed of organic-based materials, ideal substrates for microbial colonisation, are stored or exposed.

In these environments, the thermo-hygrometric parameters are quite stable, but frequently the absence of an adequate air condition systems or air ventilation may compromise the conservation of cultural assets. Moreover, the number of visitors in museums and galleries strongly influences the environmental parameters. Several environmental fungal and bacterial strains have been isolated using both passive (sedimentation method) and active sampling methods (Fig. 5.5).

For active sampling, a portable sampler (AirPort MD8, Sartorius) with a flow rate of 100 L/min, equipped with sterile disposable gelatin filter, is used (Palla 2012). Collection filters have been utilised for in vitro culture, laying it onto solid culture media, or dissolved in 1X TE solution (10 mM TRIS–HCL pH 7.5/1 mM EDTA) and utilised for the direct extraction of microbial genomic DNA and for molecular biology investigations.

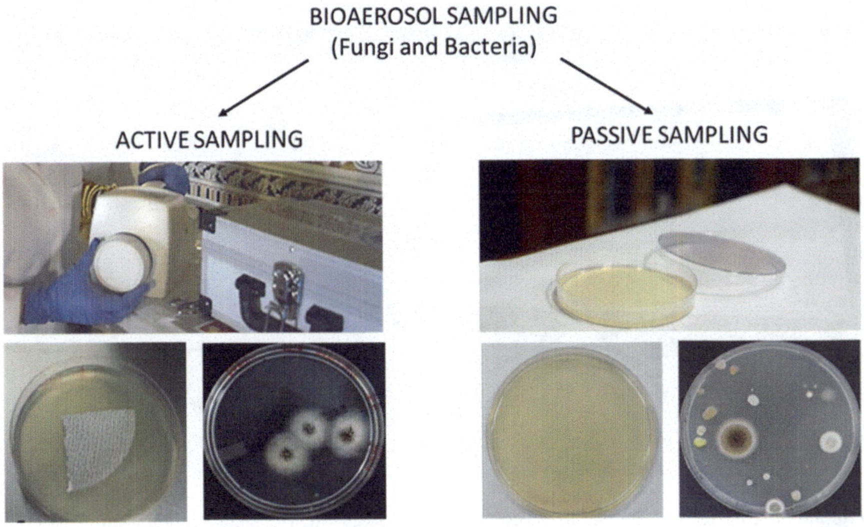

Fig. 5.5 Bioaerosol active sampling (*left*) by AirPort MD8 equipped with gelatin filters (Sartorius). The filter is layered onto nutritive media in which microbial colonies (after incubation time) have grown. Passive sampling (*right*) by gravitational sedimentation onto nutritive media, microbial growth after 16–72 h of incubation at 30 °C

5.3.1 Loop-Mediated Isothermal Amplification (LAMP) of DNA

We propose the use of a new molecular biology system, based on loop-mediated isothermal amplification (LAMP), which allows in situ genetic study without sophisticated laboratory instrument, applicable in cultural heritage field and useful to assess the presence of specific microbial taxa in the indoor environments. This technique, more sensitive and specific than qualitative PCR (Notomi et al. 2000), allows a rapid amplification of nucleic acid at constant temperature, performing tests directly on the site (Yassen et al. 2015). The proposed system (patented by Enbiotech s.r.l) is composed of a portable instrument and a set of ready to use reagents developed for this specific application. The test foresees a rapid preliminary nucleic acid extraction from the sample, genetic amplification using LAMP technology, detection of the fluorescence emitted from the sample and automatic interpretation of the results using the portable instrument (Fig. 5.6). Airborne microbiological particles are collected using AirPort MD8 air sampler, equipped with sterile gelatin filter, in order to extract the whole microbial genomic DNA, performing the LAMP test by specific primers for a chosen model pathogen microorganism.

In our hypothesis, this molecular device provides an important contribution to the development of innovative protocols to evaluate in situ the composition of microbial community, representing an improvement to PCR-based assays and microbiological analysis.

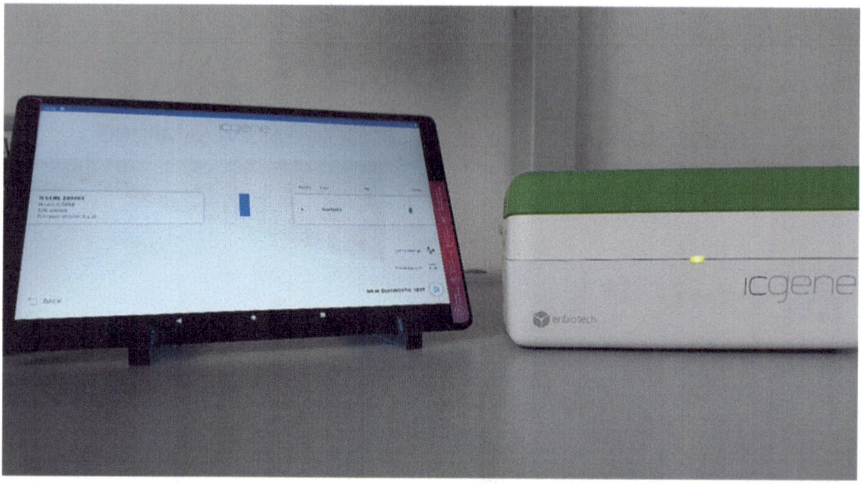

Fig. 5.6 LAMP technology. ICGENE portable mini instrument. The amplification profiles specific for bacteria or fungi cells can be directly followed on wireless connected tablet. ICGENE is a high sensitive, specific and easy-to-use in vitro diagnostic device able to perform the amplification of DNA target sequences in few minutes. The LAMP technology includes a portable fluorescent reader by measuring of light emission of analysed sample, which provides step-by-step assistance in performance of analysis by the use of an intuitive front-end (tablet and app)

5.4 Biocleaning: Bioremoval

Selected enzymes or viable bacterial cells have been applied for biological removing of undesired matter (organic layers) or crusts (sulphate, nitrate), respectively, from artwork surface. These approaches are based on the enzymatic hydrolysis of layer or on the use of deposits as carbon source or electron for viable bacterial cells (Palla et al. 2016; Alfano et al. 2011).

The scientific community and numerous technology-based start-ups are still developing the scientific knowledge base and technology to fully explore and understand the ocean and the effect of human activities. Understanding better the activities and opportunities of Blue biotechnology (Doussineau et al. 2020) is a key prerequisite for preparing for the forthcoming UN Decade of Ocean Science for Sustainable Development (2021–2030).

Blue biotechnology has the potential to address major global challenges such as sustainable food supplies, human health, energy security and environmental remediation, and to make a significant contribution to green growth in many industrial sectors (OECD 2016). Bioactive molecules was isolated from marine organisms (from fish, sponges, jellyfish, marine invertebrates, micro-algae) useful in food industry and for biomedical application (Carson and Clarke 2018).

Particularly, cold-active molecules are interesting in relation to their stability and activity at low temperature (<30 °C). Different sets of novel enzymes extracted by marine organisms with protease and esterase activity have been tested, in order to remove in a selective way undesirable or altered layers from artwork surfaces, previously characterised by high-performance liquid chromatography. The enzymatic cleaning has been performed to remove layers from artworks or ad hoc stratified laboratory specimens, applying the enzymatic gelled solutions (Klucel G, gellan gum gel, carboxymethylcellulose) in order to perform a selective and controlled action (Barresi et al. 2015; Chillè et al. 2014). Hydrolase-type enzymes are employed in the conservation of works on paper to facilitate the removal of secondary supports such as linings or mounts with advantages of specificity and efficiency in catalysing hydrolytic cleavage of polymers such as proteins, polysaccharides and lipids (Van Dyke 2004). Bioactive molecules with protease activity (BMP) have been applied to clean up protein layers from wax and papier-mâché sculptures or to remove an old *velinatura* (Japanese paper bonded with animal glue) from oil painting surfaces (Palla et al. 2015b, 2016). Moreover, animal glue layered on fresco specimen (Fig. 5.7a, b), artificially aged (for 2.200 h, UV-A 300–400 nm;

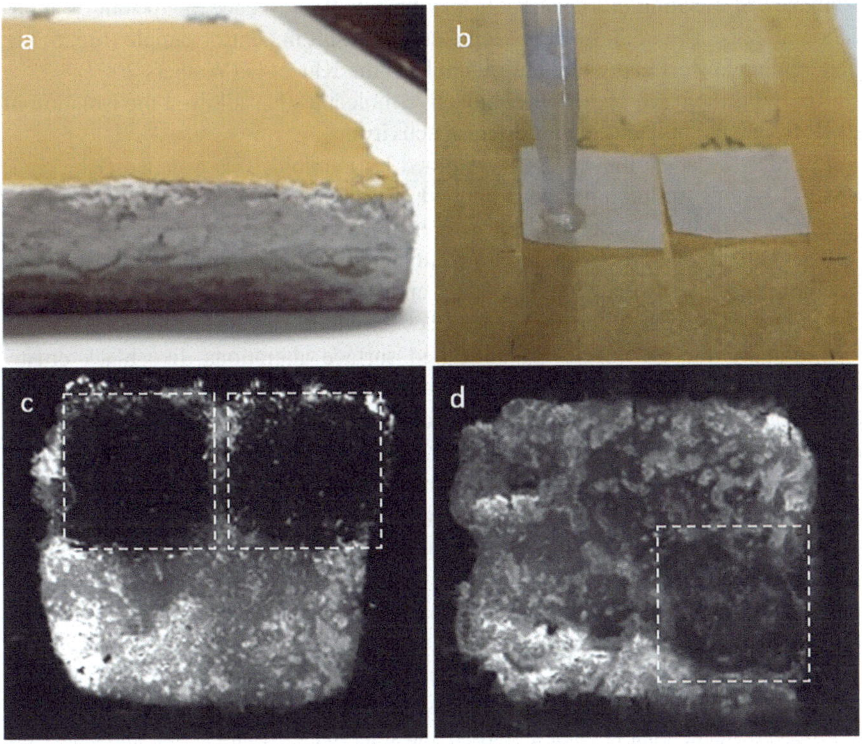

Fig. 5.7 (**a**) Fresco specimen, biocleaning tests by BMP enzyme gelled (Klucel G) solution applied, (**b**) through Japanese paper, on animal glue artificially aged layer; (**c**) efficiency of biocleaning checked by UV fluorescence irradiation by BMP, after 5′ (*left outlined area*) and 10′ (*right outlined area*); (**d**) control test (on the *bottom right*) performed by buffer gelled solution lacking BMP

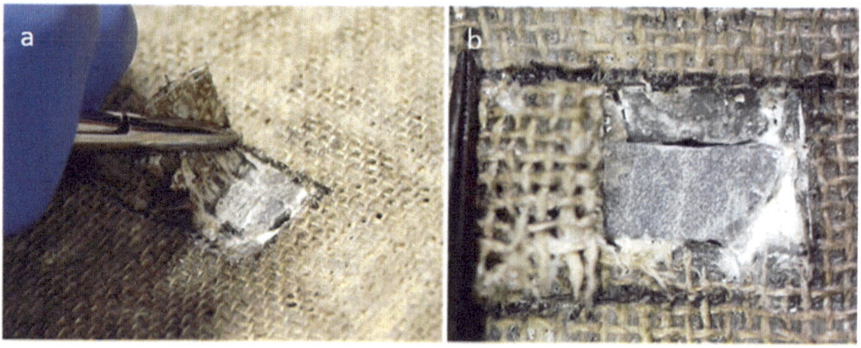

Fig. 5.8 Mosaic laboratory specimen: (**a**) canvas layer has been easily detached after 30 min of treatment; (**b**) Paraloid residues completely removed from tesserae after additional 10 min of BME application

$T = 22 \pm 5$ °C; RH = 60–65%), has been successfully hydrolysed. Particularly, the efficiency of cleaning has been confirmed by UV fluorescence, 5–10 min of application at 22 °C (Fig. 5.7c), a less deep cleaning result in control sample (Fig. 5.7d). This removal effect is probably attributable to the Klucel G (Wolbers 2007).

Recently, the research for other bioactive molecules has allowed the isolation of BME (bioactive molecules with esterase activity).

In order to detach the canvas layer glued by Paraloid B72, a gelled (Klucel G) BME (1 mg/ml) solution has been applied on canvas layer; after 30 min, the layer (naturally aged for 5 years) has been successfully removed (Fig. 5.8).

These enzymes show some important advantages, such as the possibility to work safely, at room temperatures (19–26 °C) and in a quick way (5–30 min), without heating the enzyme solutions or the surfaces on which they have been applied.

Regarding the bioremoval of undesired surface alterations, like black crusts, *Desulfovibrio vulgaris* viable cells have been particularly utilised for the removal of sulphate crusts from marble polychrome bas-relief, representing the Eternal Father exhibited at the *Regional Gallery of Palazzo Abatellis*, Palermo. The viable bacteria (*Micro4Art solfati*© produced by *Micro4you* S.r.l.) have been applied on deposit present on surface as a gelled solution. Even if the application of bacterial gel solution has been difficult, mainly due to the vertical location of the artefact and its complex geometry, the effectiveness of bioremoval of black crusts (previously characterised by XRF and FTIR analysis) was evident from the first application. Moreover, the successful removal of sulphate crusts has been confirmed by XRF analysis performed on a cleaned area (Martino et al. 2015). An important feature of this cleaning method is the speed and ease of use, and considerable importance is represented by the guarantee of a gradual and selective cleaning action and safety for the operators (non-pathogen bacteria), environment and work of art.

5.5 Microbial Growth Control

5.5.1 Bioactive Molecules from Marine Organisms

Due to the well-known antimicrobial properties of several marine organisms, the interest in application in pharmaceutical and biomedical fields (Blue Biotechnology) increases so fast.

In this scenario, bioactive molecules with antimicrobial activity (BMAs) extracted from marine invertebrates have been tested to control bacteria and fungi colonisation in laboratory specimens in order to inhibit the microbial growth on cultural assets. These BMAs have been applied in antimicrobial in vitro assays against bacteria and fungi isolated from colonised organic materials and from organic products usually used in restoration procedures. In this work, three sets of BMA molecules have been assayed on canvas and glue specimens, evaluating the control of microbial contamination (Mulè et al. 2014). In particular, the minimum inhibitory concentration (MIC) and biocide or biostatic activity has been determined by microdilution method using 96-Well Microtiter (Barresi et al. 2015). All tests have been carried out using methylparaben and benzalkonium chloride as controls.

In our hypothesis BMA molecules represent a valid alternative to traditional biocide products, safe for works of art, restores and environment, in accordance with sustainable restoration projects.

Proteases are successfully considered as an alternative to chemicals and an eco-friendly indicator for nature or the surroundings. The evolutionary relationship among acidic, neutral and alkaline proteases has been analysed based on their protein sequences, but there remains a lack of information that regulates the diversity in their specificity. Researchers are looking for microbial proteases as they can tolerate harsh conditions, ways to prevent autoproteolytic activity, stability in optimum pH and substrate specificity (Razzaq et al. 2019).

5.5.2 Plant Extracts

In the recent years, with the aim to develop alternative products to traditional biocides, several studies have been focused to various natural products with potential antimicrobial activity. Particularly, the use of plant extracts has been recognised in cultural heritage field in order to control and prevent the colonisation by microorganisms or by insect pests (Borrego et al. 2012).

Antimicrobial activity of three different plant products, *Tea tree essential oil*, *Nepeta nepetella* L. and *Allium sativum* L. extracts, has been assayed by *agar disc diffusion*, *well plate diffusion* and *microdilution* methods against bacteria and fungi, previously isolated from colonised artefacts and identified by microscopy and molecular biology investigations (Rotolo et al. 2016). Establishing for each bacteria (*Bacillus subtilis*, *Micrococcus luteus*) and fungi (*Penicillium chrysogenum*, *Aspergillus* spp.) the MIC and the inhibition halo (diameter in mm), the different susceptibility to the plant extracts has been observed during antimicrobial assays. Results

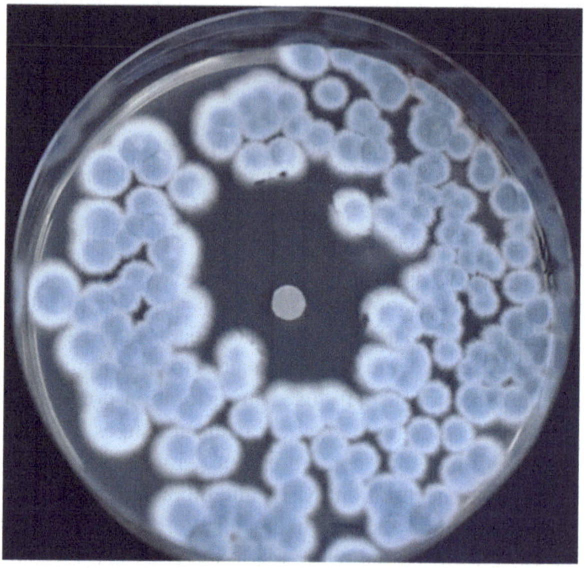

Fig. 5.9 Disc diffusion method using *Allium sativum* extract as antimicrobial agent against *Penicillium chrysogenum*, confirmed by the growth inhibition halo (Rotolo et al. 2016)

showed that the most quick method has been *agar disc diffusion* (Fig. 5.9), allowing us to hypothesise the use of these plant products as natural biocides in the control of biodeterioration of cultural assets, respecting human health and the environment (Rotolo et al. 2018; Palla et al. 2020).

5.6 Halloysite Nanotubes as Nanocarriers for Controlled Release

Inorganic nanoparticles can be employed as delivery systems for the controlled release of functional molecules useful for several applications, including medicine, pharmaceutics and cultural heritage. Compared to the synthetic nanoparticles, the nanomaterials obtained from natural sources present numerous advantages in terms of low toxicity and biocompatibility. Among the natural nanoparticles, nanoclays represent emerging materials because of their variable structural characteristics and tunable interfacial properties; halloysite nanotubes (HNTs) are very promising nanocarriers since their cavity can be easily loaded with functional molecules, such as biocides biocides (Makaremi et al. 2017; Cavallaro et al. 2018b; Stavitskaya et al. 2019; Lisuzzo et al. 2019; Vikulina et al. 2020), antioxidants (Biddeci et al. 2016; Bertolino et al. 2020; Katana et al. 2020) and antiacids (Cavallaro et al. 2017a, 2018c; Lisuzzo et al. 2021a). Interestingly, the sizes, polydispersion degree and purity of halloysite depend on its specific geological deposit (Cavallaro et al. 2018a). Tubular shape formed by aluminosilicate layers in a spiral-like morphology, is characterised by a length in the micrometer range, while the external and internal diameters are between 60–300 and 10–60 nm, respectively (Fig. 5.10). The unitary

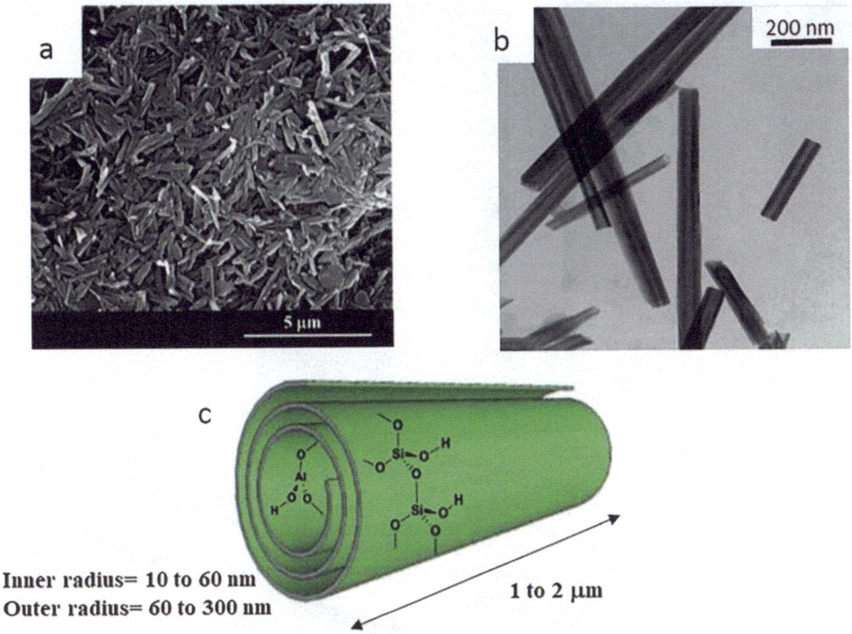

Fig. 5.10 SEM (**a**) and TEM (**b**) images of halloysite nanotubes, (**c**) sketch of hollow tubular shape of halloysite. (Cavallaro et al. 2020)

cell formula is $Al_2Si_2O_5(OH)_4{\cdot}2H_2O$, which differs to the most common kaolinite unless for the presence of the two water molecules in the interlayer.

Recently, halloysite nanotubes were investigated as nanomaterials within restoration purposes, such as consolidation and long-term protection of lignocellulosic artworks, such as archaeological woods (Cavallaro et al. 2015, 2017b; Lisuzzo et al. 2021b) and paper (Cavallaro et al. 2017a; Lisuzzo et al. 2021a) and for cleaning of marble surfaces (Lo Dico et al. 2018; Cavallaro et al. 2019). Within this, aqueous mixtures of PEG1500 and HNTs loaded with $Ca(OH)_2$ were successfully employed for the conservation of waterlogged archaeological woods. Figure 5.11 shows the optical and SEM images of wood consolidated by PEG 1500/halloysite-$Ca(OH)_2$ (percentage mass composition of 80:20).

The optical photo shows that the consolidation with PEG 1500/halloysite-Ca $(OH)_2$ conferred robustness to the archaeological wood, while SEM images highlighted that the wooden channel was successfully filled by the nanotubes (Fig. 5.11). Accordingly, we can conclude that the wood porosity decreased by the consolidation treatment with PEG 1500/halloysite-$Ca(OH)_2$ aqueous dispersion. According to the morphological investigations, the mechanical performances of the archaeological woods were significantly improved by the penetration of the consolidant within the wooden structure.

Fig. 5.11 (top) Optical photo, a mass of 100 g is placed on the top of the consolidated wood. The scale bar is 500 mm. (bottom) Scanning electron images of waterlogged archaeological wood at different magnification (**a**, **b**, **c**, **d**). The percentage mass composition of the consolidant composite was set at 80:20 for PEG 1500:halloysite-$Ca(OH)_2$. Adapted with permission from Cavallaro et al. (2020)

References

Alfano G, Lustrato G, Belli C, Zanardini E, Cappitelli F, Mello E, Sorlini C, Ranalli G (2011) The bioremoval of nitrate and sulfate alterations on artistic stonework: the case-study of Matera Cathedral after six years from the treatment. Int Biodeter Biodegradr 65(7):1004–1011. https://doi.org/10.1016/j.ibiod.2011.07.010

Altschul SF, Madden TL, Schäffer AA, Zhang J, Zhang Z, Miller W, Lipman DJ (1997) Gapped BLAST and PSI-BLAST: a new generation of protein database search programs. Nucleic Acids Res 25(17):3389–3402. https://doi.org/10.1093/nar/25.17.3389

Barresi G, Di Carlo E, Trapani MR, Parisi MG, Chillè C, Mulè MF, Cammarata M, Palla F (2015) Marine organisms as source of bioactive molecules applied in restoration projects. Herit Sci 3(3):17–20. https://doi.org/10.1186/s40494-015-0046-1

Bertolino V, Cavallaro G, Milioto S, Lazzara G (2020) Polysaccharides/Halloysite nanotubes for smart bionanocomposite materials. Carbohydr Polym 245:116502. https://doi.org/10.1016/j.carbpol.2020.116502

Biddeci G, Cavallaro G, Di Blasi F et al (2016) Halloysite nanotubes loaded with peppermint essential oil as filler for functional biopolymer film. Carbohydr Polym 152:548–557. https://doi.org/10.1016/j.carbpol.2016.07.041

Blanchette RA (2000) A review of microbial deterioration found in archaeological wood from different environments. Int Biodeterior Biodegradation 46:189–204. https://doi.org/10.1016/S0964-8305(00)00077-9

Borrego S, Valdés O, Vivar I, Lavin P, Guiamet P, Battistoni P, Gómez de Saravia SG, Borges P (2012) Essential oils of plants as biocides against microorganisms isolated from Cuban and Argentine documentary heritage. ISRN Microbiol 2012:1–7

Carson MA, Clarke SA (2018) Bioactive compounds from marine organisms: potential for bone growth and healing. Mar Drugs 16(9):340. https://doi.org/10.3390/md16090340

Cavallaro G, Lazzara G, Milioto S et al (2015) Thermal and dynamic mechanical properties of beeswax-halloysite nanocomposites for consolidating waterlogged archaeological woods. Polym Degrad Stab 120:220–225. https://doi.org/10.1016/j.polymdegradstab.2015.07.007

Cavallaro G, Danilushkina AA, Evtugyn VG et al (2017a) Halloysite nanotubes: controlled access and release by smart gates. Nano 7. https://doi.org/10.3390/nano7080199

Cavallaro G, Lazzara G, Milioto S et al (2017b) Nanocomposites based on esterified colophony and halloysite clay nanotubes as consolidants for waterlogged archaeological woods. Cellulose 24:3367–3376. https://doi.org/10.1007/s10570-017-1369-8

Cavallaro G, Chiappisi L, Pasbakhsh P et al (2018a) A structural comparison of halloysite nanotubes of different origin by small-angle neutron scattering (SANS) and electric birefringence. Appl Clay Sci 160:71–80. https://doi.org/10.1016/j.clay.2017.12.044

Cavallaro G, Lazzara G, Milioto S et al (2018b) Nanohydrogel formation within the Halloysite lumen for triggered and sustained release. ACS Appl Mater Interfaces 10:8265–8273. https://doi.org/10.1021/acsami.7b19361

Cavallaro G, Milioto S, Parisi F, Lazzara G (2018c) Halloysite nanotubes loaded with calcium hydroxide: alkaline fillers for the Deacidification of waterlogged archeological woods. ACS Appl Mater Interfaces 10:27355–27364. https://doi.org/10.1021/acsami.8b09416

Cavallaro G, Milioto S, Nigamatzyanova L et al (2019) Pickering emulsion gels based on halloysite nanotubes and ionic biopolymers: properties and cleaning action on marble surface. ACS Appl Nano Mater 2:3169–3176. https://doi.org/10.1021/acsanm.9b00487

Cavallaro G, Milioto S, Lazzara G (2020) Halloysite nanotubes: interfacial properties and applications in cultural heritage. Langmuir 36:3677–3689. https://doi.org/10.1021/acs.langmuir.0c00573

Chillè C, Di Carlo E, Barresi G, Sottile S, Lazzara G, Cammarata M, Palla F (2014) Pulitura mediante proteine enzimatiche sperimentali. In: Proceedings of XII Congresso Nazionale IGIIC - Lo Stato dell'Arte - Accademia Di Belle Arti di Brera - Milano 23/24 ottobre 2014. Nardini Editore, Firenze, Italy, pp 63–70. ISBN: 978-88-404-4451-2

Di Carlo E, Lombardo G, Barresi G, Manachini B, Palla F (2015) Biological macro and micro systems co-exist in the "Fountain of the two Dragons". Palermo Cons Sci Cult Herit 15(2): 93–105

Di Carlo E, Chisesi R, Barresi G, Barbaro S, Lombardo G, Rotolo V, Sebastianelli M, Travagliato G, Palla F (2016) Fungi and bacteria in indoor cultural heritage environments: microbial-related risks for artworks and human health. Environ Ecol Res 4(5):257–264. https://doi.org/10.13189/eer.2016.040504

Doussineau M, Haarich S, Gnamus A, Gomez J, Holstein F (2020) Smart specialisation and blue biotechnology in Europe, EUR 30521 EN, Publications Office of the European Union, Luxembourg. ISBN: 978-92-76-27753-8. https://doi.org/10.2760/19274, JRC122818

Drand J-C, Laval E (2009) X-ray fluorescence (XRF). In: Pinna D, Galeotti M, Mazzeo R (eds) Scientific examination for the investigation of paintings. A handbook for conservator-restorers. Centro Di, Firenze, pp 210–213

Katana B, Rouster P, Varga G et al (2020) Self-assembly of protamine biomacromolecule on Halloysite nanotubes for immobilization of superoxide dismutase enzyme. ACS Appl Bio Mater 3:522–530. https://doi.org/10.1021/acsabm.9b00953

Krumbein WE (2003) Patina and cultural heritage – a geomicrobiologist's prospective – how microbes change surfaces. In: Kozlowski R (ed) Cultural heritage research: a Pan European challenge, proceedings of the 5th European commission conference, Cracow, 16–18 May 2002. EUR-OP, Luxembourg, pp 1–9

Lisuzzo L, Cavallaro G, Pasbakhsh P et al (2019) Why does vacuum drive to the loading of halloysite nanotubes? The key role of water confinement. J Colloid Interface Sci 547:361–369. https://doi.org/10.1016/j.jcis.2019.04.012

Lisuzzo L, Cavallaro G, Milioto S, Lazzara G (2021a) Halloysite nanotubes filled with MgO for paper reinforcement and deacidification. Appl Clay Sci 213:106231. https://doi.org/10.1016/j.clay.2021.106231

Lisuzzo L, Hueckel T, Cavallaro G et al (2021b) Pickering emulsions based on wax and halloysite nanotubes: an ecofriendly protocol for the treatment of archeological woods. ACS Appl Mater Interfaces 13:1651–1661. https://doi.org/10.1021/acsami.0c20443

Lo Dico G, Semilia F, Milioto S et al (2018) Microemulsion encapsulated into halloysite nanotubes and their applications for cleaning of a marble surface. Appl Sci 8. https://doi.org/10.3390/app8091455

Makaremi M, Pasbakhsh P, Cavallaro G et al (2017) Effect of morphology and size of halloysite nanotubes on functional pectin bionanocomposites for food packaging applications. ACS Appl Mater Interfaces 9:17476–17488. https://doi.org/10.1021/acsami.7b04297

Martino M, Schiavone S, Balloi A, Pellegrino L, De Castro E, Palla F (2015) Bioremoval of sulphate layer from a 15th century polychrome marble artefact. Conserv Sci Cult Herit 15:235–243

Mulè MF, Di Carlo E, Barresi G, Trapani MR, Parisi MG, Cammarata M, Sottile S, Palla F (2014) Potenziali proprietà antimicrobiche di molecole bioattive: saggi su colla pasta. In: Proceedings of XII Congresso Nazionale IGIIC - Lo Stato dell'Arte - Accademia Di Belle Arti di Brera - Milano 23/24 ottobre 2014. Nardini Editore, Firenze, pp 415–422. ISBN: 978-88-404-4451-2

Notomi T, Okayama H, Masubuchi H, Yonekawa T, Watanabe K, Amino N, Hase T (2000) Loop-mediated isothermal amplification of DNA. Nucleic Acids Res 28(12):E63

OECD (2016) The ocean economy in 2030. OECD Publishing, Paris. https://doi.org/10.1787/9789264251724-en

Palla F (2009) Analisi per l'identificazione di microrganismi deteriogeni sulle superfici marmoree delle statue di Fontana Pretoria in Palermo. La Fontana Pretoria in Palermo: *hic fons, cui similis nullus in orbe patet*. Soprintendenza per i BB.CC.AA, Palermo, pp 261–266

Palla F (2012) Analytical techniques. In: Fabbri B (ed) Science and conservation for museum collections. Nardini Editore, Firenze, pp 459–470

Palla F (2015) Biological risk for the conservation and exhibition of historical–artistic artefacts in confined spaces. In: Lorusso S, Natali A, Palla F (eds) Risk management in the cultural heritage sector: museums, libraries and archives. Mimesis Edizioni, Milano-Udine, pp 95–108

Palla F (2020) Biotechnology and cultural heritage conservation. In: Turcanu-Carutiu D (ed) Heritage. London, pp 239–252

Palla F, Tartamella E (2007) Chromatic alteration on marble surfaces analysed by molecular biology tools. Conserv Sci Cult Herit 7:111–127

Palla F, Billeci N, Mancuso FP (2010) Microscopy and molecular biology techniques for the study of biocenosis diversity in semi-confined environments. Conserv Sci Cult Herit 10:185–195

Palla F, Sineo L, Manachini B (2011) Bacteria, fungi and arthropod pests collected on modern human mummies. JEAR 43(2):69–76. https://doi.org/10.4081/jear.2011.69

Palla F, Mancuso FP, Billeci N (2013) Multiple approaches to identify bacteria in archaeological waterlogged wood. J Cult Herit 14S:e61–e64. https://doi.org/10.1016/j.culher.2012.11.025

Palla F, Barresi G, Di Carlo E (2014) Identification of bacterial taxa in archaeological waterlogged wood. Conserv Sci Cult Herit 14:247–254

Palla F, Figuccio B, Sebastianelli M, Vitella M (2015a) Anthropomorphic wooden reliquaries an integrated approach to restoration. Eur J Sci Theol 11(2):25–32

Palla F, Ferrara G, Richiusa A, Teri RMC, Sebastianelli M (2015b) Application of new proteases in removal protein layers from artistic works surfaces. In: Proceedings of 7th European symposium on religious art, restoration & conservation, 4–6 June. Kermesquaderni, Nardini Editore, Trnava, pp 66–69

Palla F, Barresi G, Giordano A, Schiavone S, Trapani MR, Rotolo V, Parisi MG, Cammarata M (2016) Cold-active molecules for a sustainable preservation and restoration of historic-artistic manufacts. Int J Conserv Sci (7, Special Issue 1):239–246

Palla F, Bruno M, Mercurio F, Tantillo A, Rotolo V (2020) Essential oils as natural biocides in conservation of cultural heritage. Molecules 25(3):730. https://doi.org/10.3390/molecules25030730

Perez G, Chisesi R, Pellerito C, Pignataro B, Di Stefano C, Di Natale MC, Sebastianelli M, Palla F (2017) Scientific studies for the restoration of two wooden arm reliquaries from the Cathedral of Palermo. Eur J Sci Theol 13(2):41–53

Razzaq A, Shamsi S, Ali A, Ali Q, Sajjad M, Malik A, Ashraf M (2019) Microbial proteases applications. Front Bioeng Biotechnol 7:110. https://doi.org/10.3389/fbioe.2019.00110

Rotolo V, Barresi G, Di Carlo E, Giordano A, Lombardo G, Crimi E, Costa E, Bruno M, Palla F (2016) Plant extracts as green potential strategies to control the biodeterioration of cultural heritage. Int J Conserv Sci 7(special issue 2):839–846

Rotolo V, De Caro M, Giordano A, Palla F (2018) Solunto archaeological park in Sicily: life under mosaic tesserae. Flora Mediterr 28:233–245. https://doi.org/10.7320/FlMedit28.233

Santamaria U, Caponetti E, Caruso F, Spinella A, Fedi M, Caforio L, Tusa S, Tisseyre P, Di Stefano C, Valenti A (2009) The wood in Acqualadroni (ME-Sicily= Roman rostrum). In: Proceedings of 4th International Congress on "Science and Technology for the Safeguard of Cultural Heritage of the Mediterranean Basin", Cairo, 6th–8th December 2009, vol 2, p 53

Stavitskaya A, Batasheva S, Vinokurov V et al (2019) Antimicrobial applications of clay nanotube-based composites. Nano 9. https://doi.org/10.3390/nano9050708

Sterflinger K, Piñar G (2013) Microbial deterioration of cultural heritage and works of art - tilting at windmills? Appl Microbiol Biotechnol 97(22):9637–9646. https://doi.org/10.1007/s00253-013-5283-1

Sterflinger K, Prillinger H (2001) Molecular taxonomy and biodiversity of rock fungal communities in an urban environment (Vienna, Austria). Antonie Van Leeuwenhoek 80(3–4):275–286

Van Dyke Y (2004) Practical applications of protease enzymes in paper conservation. In: The Book and Paper Group Annual, vol 23, pp 93–107

Vikulina A, Voronin D, Fakhrullin R et al (2020) Naturally derived nano- and micro-drug delivery vehicles: halloysite, vaterite and nanocellulose. New J Chem 44:5638–5655. https://doi.org/10.1039/C9NJ06470B

Wolbers R (2007) Cleaning painted surfaces: aqueous methods. Archetype Publications, London

Yassen T, Drago S, Valentini F, Elbeaino T, Stampone G, Digiaro M, D'Onghia AM (2015) On-site detection of *Xylella fastidiosa* in host plants and in "spy insects" using the real-time loop-mediated isothermal amplification method. Phytopathol Mediterr 54(3):488–496. https://doi.org/10.14601/Phytopathol_Mediterr-15250

Taxidermal Preparations of an Extinct Sicilian Wolf Population (*Canis lupus* Linnaeus, 1758): Biological Studies and Restoration

6

Stefano Reale, Valentina Cumbo, Silvia Scibetta, Ignazio Sammarco, Sabrina Lo Brutto, Antonio Spinnato, and Salvatore Seminara

Abstract

In last decades, biotechnologies have enabled a deep development of knowledge in all fields of life sciences, allowing the collection of a wide range of data. Furthermore, the relationship between scientific investigations and conservation actions of cultural heritage is increasingly consolidated and now it is one of the peculiar points both for sustainable restoration and use of the cultural asset, also evaluating the cultural assets as source of biological information. In this study, Sicilian wolf (*Canis lupus*) specimens conserved in Sicilian Natural Museums, become a source of biological information on Sicilian biodiversity, establishing specific conservative protocol developed cooperatively with Sicilian Natural Museums. Using Next-Generation Sequencing technique, ancient DNA (aDNA) molecules extracted from nail were utilized to perform the complete mitochondrial DNA genome sequence. Phylogenetic analyses showed that this genome, which was aligned with a number of historical and extant complete wolf and dog mtDNAs sampled worldwide, was closely related to an Italian wolf mtDNA genome. These mtDNA findings suggest that wolves probably colonized Sicily from southern Italy toward the end of the last Pleistocene glacial maximum, when the Strait of Messina was almost totally dry. Furthermore, a conservative restoration of this peculiar natural specimen (taxidermal preparation) has been performed, basing on: recovery of unglued or unstitched parts; revitalization of the colors of the mucous membranes; restoration of deformed or broken parts of

S. Reale (✉) · V. Cumbo · S. Scibetta · I. Sammarco · A. Spinnato · S. Seminara
Experimental Zooprophylactic Institute of Sicily "A. Mirri", Palermo, Italy
e-mail: stefano.reale@izssicilia.it; silvia.scibetta@izssicilia.it; ignazio.sammarco@izssicilia.it; antonio.spinnato@izssicilia.it; salvatore.seminara@izssicilia.it

S. Lo Brutto
Zoological Museum "P. Doderlein", University of Palermo, Palermo, Italy
e-mail: sabrina.lobrutto@unipa.it

F. Palla, G. Barresi (eds.), *Biotechnology and Conservation of Cultural Heritage*,
https://doi.org/10.1007/978-3-030-97585-2_6

the body, washing of the coat and polishing of the fur; specific treatments against woodworms, moths, and mites; brushing and repositioning of the hair; cleaning, restoration, or creation of adequate supports or display case. The conservative intervention that should be periodically performed on each taxidermized specimen allows the protection against biodeterioration or re-infestation by parasites over time.

Keywords

Ancient DNA · mtDNA complete genome · Next-generation sequencing · Taxidermic specimen · Archeobiology

6.1 Sicilian Wolf: A Case of Insular Extinction

Biodiversity loss is one of the most dangerous threats for life on earth. The extinction of a big predator like the wolf is an impactful change for an ecological system (Mech and Boitani 2003; Paquet and Carbyn 2003; Smith et al. 2003a, b; Berger and Smith 2005; Fascione 2006); this change causes serious damage for almost all life forms of a habitat type where prey species are still living (Soulé et al. 2003). In Sicily, as in Malta or Japan, the insular wolf population were exterminated in consequence of the direct persecution as nocive species operated by humans (La Mantia and Cannella 2008; Ishiguro et al. 2009; Linnell et al. 2008), while in continental and peninsular territories wolves were able to survive keeping away from humans and its activities and then slowly demographically increase (Boitani 1984), once protection laws began to show the positive effects of wolves on protected areas and protected species, due to its predation work (Weiss et al. 2007).

Unfortunately, in these islands, where the connection with the mainland is interrupted from the last glaciation, wolves removed by humans cannot recolonize their ancient territories.

Some islands have even lost their ungulate populations and forests for anthropogenic causes but in Sicily, reforestation and the reintroduction of two wild ungulates species (*Sus scrofa* and *Dama dama*), the biocenosis are again suitable to accommodate the wolf (Leonard et al. 2005; Randi 2011; Li et al. 2016). Forests are extremely exploited and damaged by the massive impact of wild boar (Monaco et al. 2010), which tread the territory in large numbers and without the ecology of fear operated by predators (Ripple and Beschta 2004). The impact of fallow deer on tree species is still unvaluated in these territories but is believed to be considerable.

6.2 The Suggested Migrations of Wolf in Sicily

We collected five canid fossil samples from paleontological museums. At present, the analysis of one of them has been completed. Specifically, an inferior emi-mandible where the carnassial is still present and intact, which has minor dimensions than a typical modern wolf carnassial. It may be due to evolutionary reasons, as Sansalone et al. (2015) observed comparing M1 from French and Italian wolf fossils of Late and Middle Pleistocene and Holocene. They note an increase in mean value of carnassials. The partial sequence we get, compared in BLAST (NCBI) with all the sequences recorded in the platform, seems to reveal the highest percentage of identity with some Asiatic wolves, dogs and the whole sequence we get from a Sicilian wolf taxidermied in the late nineteenth century, (Aggarwal et al. 2007). We ipothize as the wolves could be migrate directory from Asia to the land that is nowadays the Sicily, and from there it expanded to Italy and Europe later, despite the most accredited (but never corroborated by studies) hypothesis about the colonization of Sicily from the the north part of Italy.

6.3 The Persecution by Humans: Last Reports of Wolves in Sicily

The wolf, present with few individuals in Sicily until the early twentieth century, was eradicated from the island due to different anthropogenic influences, such as: drastically reduction of greenwood areas; wild preys were already extinct during the previous century; absence of wild prey causing an increase in the conflict with animal husbandry; direct persecution by humans; reduction in population size such as to exacerbate the problem of hybridization with domestic dogs (La Mantia and Cannella 2008; Boitani 1984; Li et al. 2016; Hofreiter et al. 2004; Randi et al. 2014).

The forest area surveyed by the Regional Forest Inventory in Sicily (IFSR) according to the FAO definitions of "forest" and "other wooded areas," covers 512,121 hectares, compared to 101,678 hectares in 1947. Sicily, from the result of a lower number of sampling sessions compared to other regions, it has 381,647 hectares (RAF Italia, 2017–2018).

Percentage variations of the hectares of forest in Sicily: 1947–2006: increase of 259.2% in 59 years; 2006–2010: 40.2% increase in 4 years; 1947–2010: increase of 403.7% in 63 years; 2010–2019: decrease of 25.5% due to fires and deforestation.

Based on historical data (stored between 1866–1868 at sicilian naturalist Minà Palumbo museum), regarding the presence and consistency of the populations of ungulates on the island it was described as wild boar, previously abundant throughout the island, becames rare in the woods of Caronia (Messina) and Etna volcano (Catania. Areas). Moreover, he reports the deer and fallow deer as extinct and describes how rare the roe deer, previously common on Nebrodi and Madonie Sicilian mountains.

In Sicily, from 1908 to 1930, of about 1% for each species reared, 1.964,754 heads in 1908 (Checco 1983) become 1.202,514 heads in 2010 (ISTAT data), highlighting a decrease of 38.8% in the number of animals reared.

The latest reports of the killing of wolves in Sicily available in a document are collected by Fulco Pratesi (1978) in "Excluded from the ark": in 1891, San Fratello (Messina, Italy), hunters killed 7 wolves; in 1902, Santo Pietro (Caltagirone, Catania, Italy) reports the killing of a wolf; in 1924 in the Bellolampo area

(Palermo), killing of a wolf (morphology of a large dog) whose skin is still preserved at the "D'Aumale" Museum in Terrasini (Palermo, Italy), genetic analyses have confirmed that it is a dog (Angelici et al. 2019); in 1935, in the Ficuzza Forest (Palermo), the killing of a wolf is reported (documented by a photo owned by Giovanni Giardina). Minà Paumbo, in 1866–68, describes the wolf, now are present only in the woods of Madonie, Caronia, and Etna areas as a solitary animal hybridized with dogs (from morphological evaluation), therefore its extinction in nature can be traced back to the period immediately following, when certain documents on its presence are no more available. It can be considered extinct in nature as early as the end of the nineteenth century as the individuals killed in the early twentieth century were solitary and, in all probability, hybridized (the only specimen surviving from the beginning of the twentieth century was a dog, according to Angelici and collaborators in 2019), habitats and prey were no longer usual (man-made areas and domestic animals). Its alleged extinction, corresponding to the last individual killed, dates back to 1935 in Ficuzza, where the killing of a wolf is documented with photos but there is no certainty about its purity from such damaged and undefined image like it is.

6.4 The Survivors

In the world, only five specimens of wolves killed in Sicily (*Canis lupus* Linnaeus, 1758) are taxidermied and kept in Italian museums; four of them are in the Province of Palermo and one in the Natural History Museum "La Specola," in Florence. In the last years of XIX century, some scientists have recruited some individual to taxidermy preparation for museums, perhaps due to the obvious extinction in progress, so we still have, from this lost population, two prepared specimens in Zoological Museum "P. Doderlein" at the University of Palermo, an adult and a cub (in this museum there is also a skeleton of a young specimen); two other adult male are preserved in the Civic Museum "B. Romano" in Termini Imerese (Palermo) and in the Regional Museum "D'Aumale" in Terrasini (Palermo).

The four specimens of wolf houses in Sicily are objects of biomolecular analysis for the taxonomic classification of the Sicilian population.

Unfortunately, the only Sicilian wolf conserved in the Natural History Museum "La Specola" in Florence, is out of our study.

6.5 Cooperating with Museums: Restoration of Taxidermy Specimen

The four wolves kept in Sicily were in a bad state of conservation due to a lack of human resources dedicated to this mansion. We worked at the restauration of an adult male kept at the "B. Romano" Civic Museum in Termini Imerese and an adult male and a cub kept at the "P. Doderlein" Museum of Zoology, both in Palermo respectively.

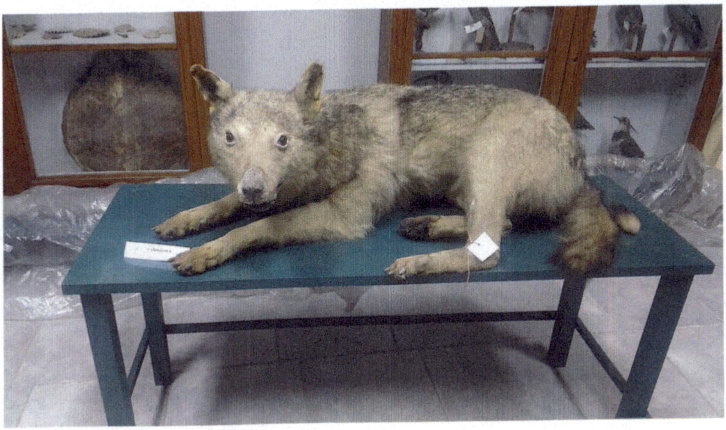

Fig. 6.1 Wolf specimen (adult male) kept at the "B. Romano" Civic Museum of Termini Imerese (Palermo, Italy); 2018 before restoration

All specimens date back to the same historical period, framed at the end of the nineteenth century, due to the techniques and materials used for the taxidermal preparations and the documents available on the specimens. Wolves went extinct in Sicily a few years later. In that sense, preserving and guaranteeing for the future the integrity of the specimens acquires greater importance and urgency.

Operations performed on the adult male wolf specimen kept at the "B.Romano" Civic Museum of Termini Imerese in July 2018 (Fig. 6.1, before restoration):

- Cleaning of the mantle was performed by a brush under a low pressure air flow, in order to remove dust and molds;
- Washing by cold water and drying with cloth and cold air;
- Brushing and repositioning of the hair according to their natural position;
- Disinfestation (pesticide treatment active against moths and mites based on tetramethrin, piperonyl butoxide, and permethrin), assembling and pesticide treatment of the wooden and plexiglass case to store the taxidermy specimen;
- Reviving the colors and the damp/wet effect of the mucous membranes, of the pads (the animal is lying on the right side with the head semi-erect), of the eyes, the rhinarium, and the claws with watercolors and fixation/polishing of the colors with vinyl glue diluted.
- Treatment of the coat with a polishing agent for skins and furs (hydrocarbons C7—n-alkanes, isoalkanes, cyclic).
- The apex of the tail, detached and kept alongside the specimen for decades, was glued with cyanoacrylate resin and kept in place for 24 h.
- The left foot had an opening in the lower part that left uncovered the iron wire of the supporting structure; this, rusty and crumbling, was removed, the internal cavity was cleaned, treated with pesticide, filled with cotton wool, and sutured with cyanoacrylate resin. The claw of the fourth toe of the left foot was

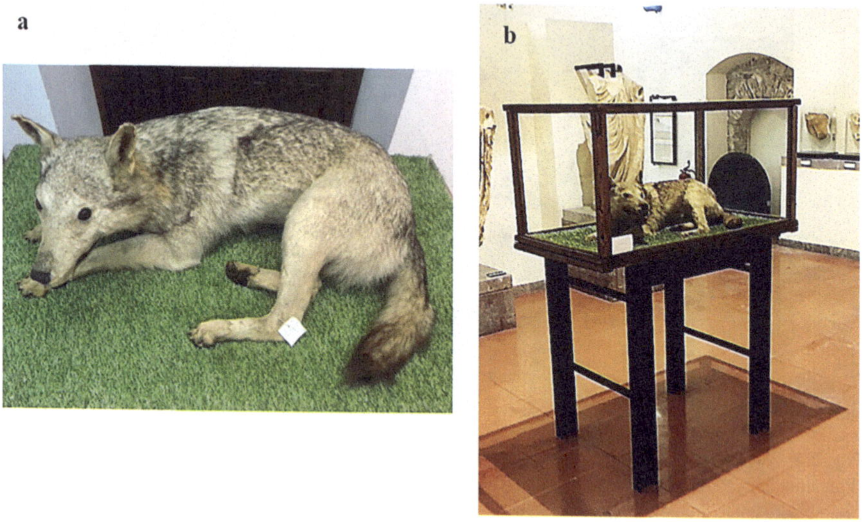

Fig. 6.2 Wolf specimen kept at the "B. Romano" Civic Museum of Termini Imerese (Palermo, Italy): (**a**) after restoration; (**b**) exposed in a wood and plexiglass case

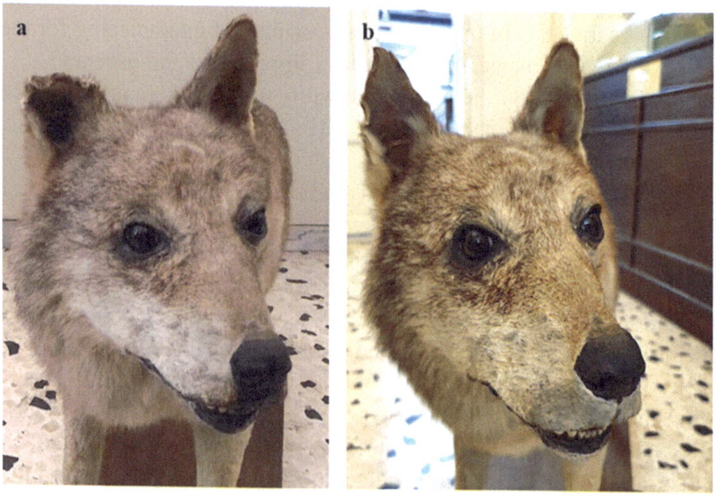

Fig. 6.3 Wolf specimen kept at the "P. Doderlein" Museum of Zoology at the University of Palermo: (**a**) before and (**b**) after restoration

reconstructed following the taking of the biological sample for biomolecular analyses (Fig. 6.2a).

Operations performed on the adult male wolf specimen at the "P. Doderlein" Zoology Museum (PA), June 2020, (Fig. 6.3, before restoration):

- Cleaning of the mantle performed with a brush and low pressure compressed air for the elimination of dust and the removal of molds;
- Washing by cold water and drying with cloth and cold air;
- Brushing and repositioning of the hair according to their natural position;
- Disinfestation (pesticide treatment active against moths and mites, based on tetramethrin, piperonyl butoxide, and permethrin) of the taxidermy specimen and of the wooden support;
- Reviving the colors and the damp/wet effect of the mucous membranes, pads, eyes, rhinarium, and claws with watercolors and fixing/polishing the colors with diluted vinyl glue;
- Treatment of the coat with a polishing agent for skins and furs (hydrocarbons C7—n-alkanes, isoalkanes, cyclic);
- The animal's right ear was bent forward due to incorrect positioning inside the display case, it was softened with water, straightened and held in pose with a brace for 24 h;
- Cleaning the wooden support;
- To take the biological sample intended for biomolecular analyses, the animal was removed from the support (it is in an upright position, fixed on the wooden platform by the ends of the wire protruding from the pads and forming part of the supporting structure of the taxidermal preparation); the innermost parts of the limbs (mainly bone tissue) were removed using a small drill for jewelry, starting from the exit hole between the pads. The holes made were rebuilt with epoxy resin and colored before reassembling the specimen on its support.

Operations performed on the young wolf specimen at the "P. Doderlein" Zoological Museum, June 2020 (Fig. 6.3a, before restoration):

- Cleaning of the mantle performed with a brush and low pressure compressed air for the elimination of dust and the removal of molds;
- Washing with cold water and drying with cloth and cold air;
- Brushing and repositioning of the hair according to their natural position;
- Disinfestation (pesticide treatment active against moths and mites based on tetramethrin, piperonyl butoxide, and permethrin) of the taxidermy specimen and of the wooden support;
- Reviving the colors and the damp/wet effect of the mucous membranes, pads, eyes, the rhinarium, and claws with watercolors and fixing/polishing the colors with diluted vinyl glue;
- Treatment of the coat with a polishing agent for skins and furs (hydrocarbons C7—n-alkanes, isoalkanes, cyclic);
- Cleaning the wooden support;
- Evident is the improvement in conservation of this peculiar cultural heritage (Fig. 6.3b).

To take the biological sample intended for biomolecular analyses, the animal was removed from the support (it is in an upright position, fixed on the wooden platform by the ends of the wire protruding from the pads and forming part of the supporting structure of the taxidermal preparation); the innermost parts of the limbs (mainly bone tissue) were removed using a small drill for jewelry, starting from the exit hole between the pads. The holes made were rebuilt with epoxy resin and colored before reassembling the specimen on its support.

6.5.1 Cooperating with Museums: DNA Shots Caught from the Past

Museums constitute a challenge for zoologists who work on taxonomy and biogeography of species, subspecies, and populations. Natural and paleontological museums are an interesting and amazing playground for those who want to reconstruct the natural history of a territory and its faunal migrations, colonizations, and disappearance.

These kinds of collaboration are those that increase knowledge and prestige of both entities—research institutes and museums—and that is something we believe is welcome to regularly happen.

6.6 Biomolecular Analysis

The decomposition of organic matter often does not affect the presence of trace amounts of DNA, even thousand years after the death of the animal. These amount are usually fragmented due to depurination (Lindahl 1993; Briggs et al. 2007). The correlation between the degree of DNA fragmentation and the sample age is valid for the same site (Allentoft et al. 2012) but not for different sites (Sawyer et al. 2012). Temperature and other environmental conditions affect DNA preservation (Smith et al. 2003a, b; Shapiro et al. 2004).

Van der Valk et al. (2021) have recently pushed the molecular analysis to a million years ago, finding two distinct mammoth lineages in eastern Siberia during the Early Pleistocene.

6.6.1 DNA Extraction

Ancient DNA (aDNA) research and the field of paleogenomics (i.e., genome-wide aDNA analysis of humans as well as animals, plants, and other organisms) offer opportunities to understand the actual world and evolution in stimulating ways (Shapiro and Hofreiter 2014; MacHugh et al. 2017; Krings et al. 1997). The specimen preserved at the Civic Museum in Termini Imerese (Palermo) was sampled by taking the inner part of the phalanx and the nail to prevent contamination and DNA degradation. The sample was transferred to a UV-decontaminated test tube.

Bone tissues obtained from the specimen kept in "D'Aumale" Museum in Terrasini (Palermo) were processed with Promega Bone Incubation Buffer and then extracted with Tissue and Hair Extraction Kit (for use with DNA IQ™,

Promega Corporation, Madison, USA). This kit is dedicated to formalin-fixed and paraffin-embedded tissue: these substances can compromise the fragmentation and amplification of DNA by binding functional groups for chemical reactions. Rates obtained from inner parts of fossils records were processed with ChargeSwitch® Forensic DNA Purification Kit (Thermo Fisher) without adding Proteinase-K in the lysis mix, because of the absence of proteins already degraded during millennials and fossilization processes which substitute organic components with carbonates.

It recommends working in a DNA-free-zone of the lab (regularly decontaminated), where DNA derived from the same genus was never processed before and nothing else has to be processed the same day. Nothing else was processed in the lab those days. We used a separate area of the lab dedicated to special applications, easy to decontaminate and that was never employed for Canis spp. DNA. All bench-tops and equipment were flamed or cleaned with bleach and ethanol.

We used a 1% solution of sodium hypochlorite, sprayed on the lab surfaces and left for 5 min before drying and wiped with 70% ethanol to produce levels of gaseous chlorine at the recommended exposure limits (Ballantyne et al. 2015).

All the equipment were exposed to UV overnight in a sterile laminar-flow hood. We used pipette tips with sterile aerosol filters.

6.6.2 Library Preparation for NGS Sequencing

Spectrophotometer analysis was used to check quality and concentration of the DNA using Nanodrop (Thermo Fisher Scientific, Waltham, Massachusetts, USA), Qubit (Thermo Fisher Scientific, Waltham, Massachusetts, USA), and TapeStation (Agilent Technologies Inc. Santa Clara, California) instruments. Negative sample used for blank was water and no DNA trace was found in it. Fragments smaller than 200 bp were removed by Agencourt (Beckman) beads protocol. We used the Illumina TruSeq DNA nano kit for the specimen of the Civic Museum in Termini Imerese and Nextera DNA Library Prep (Illumina Inc., San Diego, CA, USA) for library preparation of fossils records and the specimen preserved in "D'Aumale" Museum in Terrasini, Palermo. The shortest DNA fragments were removed with 250 l of undiluted Sample Purification Beads (SPB) reagent.

Extracted DNA from fossils and "Terrasini" specimens was unified sample by sample and concentrated with Vivacon® 500 (Sartorius Stedim Lab Ltd., Stonehouse, UK) procedure using the 50,000 membrane to select fragments up to 300 bp.

PhiX Control library (v2; Illumina) was added. The "Termini" library was sequenced on a MiSeq platform (Illumina Inc., San Diego, CA, USA), using a SBS MiSeq Reagent Kit v2 with a 150 Paired End run. Fossils records and "D'Aumale" specimens (6 libraries) were sequenced on HiSeq 2500 platform (Illumina Inc., San Diego, CA, USA).

6.6.3 Checking DNA Quality and Species Belonging

Complete mitochondrial genome sequence was obtained for seven specimens, from one of them but was reconverted in Fastq reads. The reads were mapped using the complete mitochondrial genome of an Italian wolf recorded in GenBank (KF661048) as reference by means of the BWA-MEM 0.7.17 aligner algorithm (Li and Durbin 2010). The genome obtained was mapped using the reference mitogenome and filtered with MapDamage 2.0 (Jónsson et al. 2013): C->T or G->A transitions, potentially due to DNA degradation, were modified according to the position in reads and damage models. Filtered reads were selected from the BAM file using SAMTools 1.4 (Li et al. 2009) by converting them to Fastq format and assembled into a complete consensus sequence with Spades 3.11 (Bankevich et al. 2012) for phylogenetic analysis. The quality of assembled mtDNA genome was verified using Quast tool (Gurevich et al. 2013). The DNA SNPs were checked by manual inspection using IGV 2.4.6.

Our new SRY sequence resulted identical to *Canis lupus* familiaris chromosome Y genomic sequence, accession number KP081776.1, with a 100% of identity and an e-value of 5e−49. The wolf mitogenome we obtained was aligned with the sequences of the three Italian wolf mtDNA genomes in GenBank: KF661048.1 (Thalmann et al. 2013); KU696389.1 and KU644662.1 (Koblmuller et al. 2016) and with complete Canis lupus mitogenomes in the GenBank using ClustalW (Higgins et al. 1994) in MEGAX (Kumar et al. 2018) with Neighbor-joining algorithm. Translations were considered and short overlapping fragments, incomplete codons and mutations were identified working on MEGAX. We excluded missing or ambiguous bases from our analysis.

6.7 Phylogenetic Analysis

Two different datasets obtained aligning the new Sicilian and other *Canis lupus* mitogenomes downloaded from GenBank were analyzed, regarding "Termini Imerese" specimens mtDNA genome (Reale et al. 2019). The first one includes 51 complete modern wolf mitogenomes used by Koblmuller et al. (2016), three historical wolf samples, 14 dogs (*C.l.familiaris*) and five Himalayan wolves (named *C.l.laniger* or *C.l.chanco*), used as outgroups. The second one includes 26 modern wolf mtDNA used by Thalmann et al. (2013), 2 historical Ezo wolf (*C.l.hattai*) and 3 historical Honsu wolf (*C.l.hodophilax*; Matsumura et al. 2014), 4 Himalayan wolves and 4 coyote used as outgroups. For alignments, control regions (CRs) were excluded because of the absence of this region in historical and some modern wolves. Software such as MEGAX were employed for manual alignments. Sicilian CR was loaded in BLAST to be compared to that of wolves and dogs present in the platform dataset. The two sets of mtDNA chosen were compared and aligned in MEGAX using the FindBest DNA Model maximum likelihood (ML) procedure. The Neighbor-joining procedure (NJ; Saitou and Nei 1987) was the best model to use. All positions containing missing nucleotides were excluded. One set of alignments was used to compute a Bayesian majority-rule phylogenetic tree using BEAST 2.5.1 (Drummond et al. 2012), with the HKY + G model. Markov chain Monte Carlo (MCMC) samples were drawn every 1000 generations from a total of 1,000,000 generations, following a discarded burn-in, based on Tracer 1.7.1 (Drummond and Rambaut 2007) outputs, of 100,000 generations. BEAST was used to calculate the age of the most recent common ancestor (MRCA) of a clade using the Italian and Sicilian wolf mitogenome, calibrated with the divergence times among Japanese wolves (MRCA = 46,800; 95% highest probability density HPD = 37,500–58,000 years) by Matsumura et al. (2014) sensu. Both the uncorrelated log-linear model (Drummond et al. 2006) and the strict clock model were tested for the molecular clock, and both the Bayesian skygrid model (Gill et al. 2013) and a constant model were used for the population size. Tracer 1.7.1. was used to assess the convergence and performance of different models. Tree 1.4.2. and TreeAnnotator in BEAST performed the consensus tree (Rambaut 2014). With BEAUti (in the BEAST package) was developed the configuration file for BEAST.

All fragments that compose the whole mtDNA genome (16,678 bp) were obtained assembling the reads generated by NGS. This new mtDNA wolf genome was submitted and registered with an accession number MH891616 in GenBank. In MEGAX the alignment with reference dog and Italian wolf homologous sequences, by direct inspection, proved that the mapping and the identification of all the expected mtDNA tRNA, rRNA, and protein-coding genes were correct; these sequences did not show any anomalous stop codon and translated into the expected RNAs or proteins. Thus, was assumed that this mtDNA genome was authentic. The Sicilian wolf mtDNA joined a basal phylogenetic clade that included the Italian wolf and two mtDNA genomes, respectively, sequenced from a wolf sampled in Poland (KF661045) and in Belarus (KU696390) in both trees. This clade (the Italian clade) was basal to all the other modern wolf and dog genomes, excluding two Pleistocene wolves: KF661088 (estimated age 28,000 years, from Alaska) and KF661081 (18,000 years from Russia). The Sicilian wolf mtDNA in the Italian clade turned out to be basal to the peninsular Italian wolf mtDNA. The Italian clade was 99%–100% supported by bootstrap and interior-branch length tests. The same results were obtained in MEGAX with the best-fit TN93 + G substitution model and the NJ procedure.

The Japanese and Italian wolf clades were basal to all the other modern wolf and dog genomes in the Bayesian majority-rule consensus phylogenetic trees obtained with 4 coyote (*C. latrans*) mtDNAs used as outgroups, supporting the IQtree and MEGAX results. Was estimated by BEAST software, following Matsumura et al. (2014) to calculate the average coalescence time of the Italian clade (MRCA = 13,400 years, with 95% highest probability density HPD = 4000–21,230). Similar results were observed in the divergence times of the Italian wolf and Japanese wolf from the other wolf clades (c. 100,000 years), highlighting the ancient origins of wolves in peninsular Italy and Sicily in comparison to other wolves and dogs worldwide. The Italian and Sicilian wolf mtDNA belong to the wolf haplogroup-2 that includes all the ancient wolves sampled in western Europe dating from between 44,000 and 1200 years ago (Pilot et al. 2010). The CR of the Sicilian wolf did not match with any of the dog sequences in BLAST (GenBank Tool), thus supporting its origin in a wild wolf population. The CR sequence was identical to a partial mtDNA CR sequence found by Dufresnes et al. (2018) in a different wolf sample from Sicily (the skeleton of a young wolf kept at "P. Doderlein Museum"). The sequence was also identical to two partial mtDNA CR sequences (MK129178 and MK129179) found by Angelici et al. (2019). Consequently, this CR seems to be unique for the extinct wolf population of Sicily. We also searched the stored genomic DNA reads for specific sex markers. We found that SRY sequences matching with the homologous *Canis lupus* chromosome Y genomic sequences present in GenBank (AF107021), indicating that the specimen in study was a male.

For the first time a complete mitogenome of a wolf from the extinct Sicilian population was available (Reale et al. 2019).

6.8 A Future Perspective for Sicilian Wolf Population

Paleogenomics gives us a new instrument of understanding paleoecological dynamics and paleoenvironmental conditions through aDNA, which can survive much more than ever known (Dabney et al. 2013).

The divergence time of 13,400 years (95% HPD = 4000–21,230) of the Sicilian wolf mitogenome overlaps with the age of the last land bridge between the main island and the Italian south-western territory lasted for at least 500 year from 21,500 to 20,000 years ago (Antonioli et al. 2014).

The planet definitely loses an unique representative population of wolves that show an ancient haplotype, different from any other extinct or vivant wolf population (Reale et al. 2019). This loss is irrecoverable, in the light of the biodiversity value, and nowadays, with the presence of introduced populations of wild boar and fallow deer in two different Sicilian mountain masses and an important reforestation (both spontaneous and artificial) of the territory, the wolf presence could be newly pertinent, in an ecological view. But, at the present, the ancestral human attitude against wolves, the presence of feral dogs and the possible impact on farm activities can compromise the colonization of a new wolf population, because of direct killing of problematic individuals.

The Italian wolf is the most close and related to the extinct Sicilian wolf (Reale et al. 2019) and it could be prefigured as the best candidate for a reintroduction, even considering the demographic expansion of the peninsular population (Fabbri et al. 2007; Mattioli et al. 2014; LIFE WOLFALPS Final Report 2018).

References

Aggarwal RK, Kivisild T, Ramadevi J, Sing L (2007) Mitochondrial DNA coding region sequences support the phylogenetic distinction of two Indian wolf species. J Zool Syst Evol Res 45:163–172. https://doi.org/10.1111/j.1439-0469.2006.00400.x

Allentoft ME, Collins M, Harker D, Haile J, Oskam CL, Hale ML, Campos PF, Samaniego JA, Gilbert MTP, Willerslev E, Zhang G, Scofield RP, Holdaway RN, Bunce M (2012) The half-life of DNA in bone: measuring decay kinetics in 158 dated fossils. Proc Biol Sci B 279(1748): 4724–4733. https://doi.org/10.1098/rspb.2012.1745

Angelici FM, Ciucani MM, Angelini S, Annesi F, Caniglia R, Castiglia R, Fabbri E, Galaverni M, Palumbo D, Ravegnini G, Rossi L, Siracusa AM, Cilli E (2019) The Sicilian wolf: genetic identity of a recently extinct insular population. Zool Sci 36(3):189–197. https://doi.org/10.2108/zs180180

Antonioli F, Lo Presti V, Morticelli MG, Bonfiglio L, Mannino MA, Palombo MR, Tonielli R (2014) Timing of the emergence of the Europe–Sicily bridge (40–17 cal ka BP) and its implications for the spread of modern humans. Geol Soc Lond, Spec Publ 411(1):111–144. https://doi.org/10.1144/SP411.1

Ballantyne KN, Salemi R, Guarino F, Pearson JR, Garlepp D, Fowler S, vanOorschot RAH (2015) DNA contamination minimisation – finding an effective cleaning method. Aust J Forensic Sci 47(4):428–439. https://doi.org/10.1080/00450618.2015.1004195

Bankevich A, Nurk S, Antipov D, Gurevich AA, Dvorkin M, Kulikov AS, Lesin VM, Nikolenko SI, Pham S, Prjibelski AD, Pyshkin AV, Sirotkin AV, Vyahhi N, Tesler G, Alekseyev MA, Pevzner PA (2012) SPAdes: a new genome assembly algorithm and its applications to single-cell sequencing. J Comput Biol 19(5):455–477. https://doi.org/10.1089/cmb.2012.0021

Berger J, Smith DW (2005) Restoring functionality in Yellowstone with recovering carnivores: gains and uncertainties. In: Ray JC, Redford KH, Steneck RS, Berger J (eds) Large carnivores and the conservation of biodiversity. Island Press, pp 100–109

Boitani L (1984) Genetic considerations on wolf conservation in Italy. Boll Zool 51:367–373. https://doi.org/10.1080/11250008409439476

Briggs AW, Stenzel U, Johnson PLF, Green RE, Kelso J, Prüfer K, Meyer M, Krause J, Ronan MT, Lachmann M, Pääbo S (2007) Patterns of damage in genomic DNA sequences from a Neandertal. Proc Natl Acad Sci U S A 104(37):14616–14621. https://doi.org/10.1073/pnas.0704665104

Checco A (1983) Banca e latifondo della Sicilia negli anni Trenta. Istituto siciliano per la storia contemporanea, Aloisi

Dabney J, Knapp M, Glocke I, Gansauge MT, Weihmann A, Nickel B, Valdiosera C, García N, Pääbo S, Arsuaga JL, Meyer M (2013) Complete mitochondrial genome sequence of a Middle Pleistocene cave bear reconstructed from ultrashort DNA fragments. PNAS USA 110:15758–15763. https://doi.org/10.1073/pnas.1314445110

Drummond AJ, Rambaut A (2007) BEAST: Bayesian evolutionary analysis by sampling trees. BMC Evol Biol 7:214. https://doi.org/10.1186/1471-2148-7-214

Drummond AJ, Ho SY, Phillips MJ, Rambaut A (2006) Relaxed phylogenetics and dating with confidence. PLoS Biol 4(5):e88. https://doi.org/10.1371/journal.pbio.0040088

Drummond AJ, Suchard MA, Xie D, Rambaut A (2012) Bayesian phylogenetics with BEAUti and the BEAST 1.7. Mol Biol Evol 29(8):1969–1973. https://doi.org/10.1093/molbev/mss075

Dufresnes C, Miquel C, Remollino N, Biollaz SN, Taberlet P, Fumagalli L (2018) Howling from the past: historical phylogeography and diversity losses in European grey wolves. Proc Biol Sci B 285. https://doi.org/10.1098/rspb.2018.1148

Fabbri E, Miquel C, Lucchini V, Santini A, Caniglia R, Duchamp C, Weber JM, Lequette B, Marucco F, Boitani L, Fumagalli L, Taberlet P, Randi E (2007) From the Apennines to the Alps: colonization genetics of the naturally expanding Italian wolf (*Canis lupus*) population. Mol Ecol 16(8):1661–1671. https://doi.org/10.1111/j.1365-294X.2007.03262.x

Fascione N (2006) Places for wolves: a blueprint for restoration and recovery in the lower 48 states. Defenders of Wildlife

Gill MS, Lemey P, Nuno R, Faria NR, Rambaut A, Shapiro B, Suchard MA (2013) Improving Bayesian population dynamics inference: a coalescent-based model for multiple loci. Mol Biol Evol 30(3):713–724. https://doi.org/10.1093/molbev/mss265

Gurevich A, Saveliev V, Vyahhi N, Tesler G (2013) QUAST: quality assessment tool for genome assemblies. Bioinformatics 29(8):1072–1075. https://doi.org/10.1093/bioinformatics/btt086

Higgins D, Thompson J, Gibson T, Thompson JD, Higgins DG, Gibson TJ (1994) CLUSTAL W: improving the sensitivity of progressive multiple sequence alignment through sequence weighting, position-specific gap penalties and weight matrix choice. Nucleic Acids Res 22(22):4673–4680. https://doi.org/10.1093/nar/22.22.4673

Hofreiter M, Serre D, Rohland N, Rabeder G, Nagel D, Conard N, Munzel S, Pääbo S (2004) Lack of phylogeography in European mammals before the last glaciation. PNAS USA 101:12963–12968. https://doi.org/10.1073/pnas.0403618101

Ishiguro N, Inoshim Y, Shigehara N (2009) Mitochondrial DNA analysis of the Japanese wolf (*Canis lupus hodophilax* Temminck, 1839) and comparison with representative wolf and domestic dog haplotypes. Zool Sci 26(11):765–770. https://doi.org/10.2108/zsj.26.765

Jónsson H, Ginolhac A, Schubert M, Johnson P, Orlando L (2013) MapDamage2.0: fast approximate Bayesian estimates of ancient DNA damage parameters. Bioinformatics 29(13): 1682–1684. https://doi.org/10.1093/bioinformatics/btt193

Koblmuller S, Vilà C, Lorente-Galdos B, Dabad M, Ramirez O, Marques-Bonet T, Wayne RK, Leonard JA (2016) Whole mitochondrial genomes illuminate ancient intercontinental dispersals of grey wolves (*Canis lupus*). J Biogeogr 43:1728–1738. https://doi.org/10.1111/jbi.12765

Krings M, Stone A, Schmitz RW, Krainitzki H, Stoneking M, Pääbo S (1997) Neandertal DNA sequences and the origin of modern humans. Cell 90(1):19–30. https://doi.org/10.1016/s0092-8674(00)80310-4

Kumar S, Stecher G, Li M, Knyaz C, Tamura K (2018) MEGA X: molecular evolutionary genetics analysis across computing platforms. Mol Biol Evol 35(6):1547–1549. https://doi.org/10.1093/molbev/msy096

La Mantia T, Cannella Z (2008) Presenza storica dei grossi mammiferi in sicilia. In: 87–112 Atlante Della Biodiversità Della Sicilia: Vertebrati Terrestri. Studi e Ricerche 6, Arpa Sicilia, Palermo, pp 536

Leonard J, Vilà C, Wayne RK (2005) Legacy lost: genetic variability and population size of extirpated US grey wolves (*Canis lupus*). Mol Ecol 14(1):9–17. https://doi.org/10.1111/j.1365-294X.2004.02389.x

Li H, Durbin R (2010) Fast and accurate long-read alignment with burrows-wheeler transform. Bioinformatics 26(5):589–595. https://doi.org/10.1093/bioinformatics/btp698

Li H, Handsaker B, Wysoker A, Fennell T, Ruan J, Homer N, Marth G, Abecasis G, Durbin R (2009) The sequence alignment/map format and SAMtools. Bioinformatics 25(16):2078–2079. https://doi.org/10.1093/bioinformatics/btp352

Li H, Xiang-Yub J, Daib G, Gub Z, Ming C, Yang Z, Ryder OA, Li W-H, FuY-X ZY-P (2016) Large numbers of vertebrates began rapid population decline in the late 19th century. PNAS 113(49):14079–14084. https://doi.org/10.1073/pnas.1616804113

LIFE12 NAT/IT/000807 LIFE WOLFALPS – FINAL REPORT (2018)

Lindahl T (1993) Instability and decay of the primary structure of DNA. Nature 362(6422): 709–715. https://doi.org/10.1038/362709a0

Linnell J, Salvatori V, Boitani L (2008) Guidelines for population level management plans for large carnivores in Europe. In: A Large Carnivore Initiative for Europe Report Prepared for the European Commission (contract 070501/2005/424162/MAR/B2)

MacHugh DE, Larson G, Orlando L (2017) Taming the past: ancient DNA and the study of animal domestication. Annu Rev Anim Biosci 5:329–351. https://doi.org/10.1146/annurev-animal-022516-022747

Matsumura S, Inoshima Y, Ishiguro N (2014) Reconstructing the colonization history of lost wolf lineages by the analysis of the mitochondrial genome. Mol Phylogenet Evol 80(1):105–112. https://doi.org/10.1016/j.ympev.2014.08.004

Mattioli L, Forconi P, Berzi D, Perco F (2014) Wolf population estimate in Italy and monitoring perspectives. Conference: IX Congresso italiano di Teriologia Civitella Alfedena (AQ), 7-10 maggio 2014. At: Hystrix the Italian Journal of Mammalogy. Volume: 25 (Supplement)

Mech LD, Boitani L (eds) (2003) Wolves: behavior, ecology, and conservation. University of Chicago Press

Minà Palumbo F (1868) Catalogo dei mammiferi della Sicilia. Ann Agricoltura Siciliana 12(Serie 2):3–123

Monaco A, Carnevali L, Toso S (2010) Linee guida per la gestione del Cinghiale (*Sus scrofa*) nelle aree protette. 2° edizione. Quad Cons Natura 34, Min.Ambiente – ISPRA

Paquet PC, Carbyn LN (2003) Gray wolf. In: Feldhamer GA, Thompson BC, Chapman JA (eds) Wild mammals of North America: biology, management, and conservation, 2nd edn. John Hopkins University Press, pp 482–510

Pilot M, Branicki W, Jedrzejewsk W, Goszczynski J, Jedrzejewska B, Dykyy I, Shkvyrya M, Tsingarska E (2010) Phylogeographic history of grey wolves in Europe. BMC Evol Biol 10. https://doi.org/10.1186/1471-2148-10-104

Pratesi F (1978) Esclusi dall'arca: animali estinti e in via di estinzione in Italia. Arnoldo Mondadori Editore Spa, Milano, pp 32–43

Rambaut A (2014) FigTree v1.4.2, a graphical viewer of phylogenetic trees. Available from http://tree.bio.ed.ac.uk/software/figtree/

Randi E (2011) Genetics and conservation of wolves *Canis lupus* in Europe. Mammal Rev 41(2): 99–111. https://doi.org/10.1111/j.1365-2907.2010.00176.x

Randi E, Hulva P, Fabbri E, Galaverni M, Galov A, Kusak J, Bigi D, Cerná Bolfíková B, Smetanová M, Caniglia R (2014) Multilocus detection of wolf x dog hybridization in Italy, and guidelines for marker selection. PLoS One 9(3):e91412. https://doi.org/10.1371/journal.pone.0086409

Rapporto sullo stato delle foreste e del settore forestale in Italia, 2017–2018. Compagnia delle Foreste S.r.l.

Reale S, Randi E, Cumbo V, Sammarco I, Bonanno F, Spinnato A, Seminara S (2019) Biodiversity lost: the phylogenetic relationships of a complete mitochondrial DNA genome sequenced from the extinct wolf population of Sicily. Mamm Biol 98:1–10. https://doi.org/10.1016/j.mambio.2019.06.002

Ripple WJ, Beschta RL (2004) Wolves, elk, willows, and trophic cascades in the upper Gallatin range of southwestern Montana, USA. For Ecol Manag 200(1–3):161–181. https://doi.org/10.1016/j.foreco.2004.06.017

Saitou N, Nei M (1987) The neighbor-joining method: a new method for reconstructing phylogenetic trees. Mol Biol Evol 4(4):406–425. https://doi.org/10.1093/oxfordjournals.molbev.a040454

Sansalone G, Bertè DF, Maiorino L, Pandolfi L (2015) Evolutionary trends and stasis in carnassial teeth of European Pleistocene wolf *Canis lupus* (*Mammalia*, *Canidae*). Q Sci Rev 110:36–48. ISSN: 0277-3791

Sawyer S, Krause J, Guschanski K, Savolainen V, Pääbo S (2012) Temporal patterns of nucleotide misincorporations and DNA fragmentation in ancient DNA. PLoS One 7(3):e34131. https://doi.org/10.1371/journal.pone.0034131

Shapiro B, Hofreiter M (2014) A paleogenomic perspective on evolution and gene function: new insights from ancient DNA. Science 343(6169):1236573. https://doi.org/10.1126/science.1236573

Shapiro B, Drummond AJ, Rambaut A, Wilson MC, Matheus PE, Sher AV, Pybus OG, Gilbert MTP, Barnes I, Binladen J, Willerslev E, Hansen AJ, Baryshnikov GF, Burns JA, Davydov S, Driver JC, Froese DG, Harington R, Keddie G, Kosintsev P, Kunz ML, Martin LD, Stephenson RO, Storer J, Tedford R, Zimov S, Cooper A (2004) Rise and fall of the Beringian steppe bison. Science 306(5701):1561–1565. https://doi.org/10.1126/science.1101074

Smith DW, Peterson RO, Houston DB (2003a) Yellowstone after wolves. Bioscience 53(4): 330–340. https://doi.org/10.1641/0006-3568(2003)053[0330:YAW]2.0.CO;2

Smith CI, Chamberlain AT, Riley MS, Stringer C, Collins MJ (2003b) The thermal history of human fossils and the likelihood of successful DNA amplification. J Hum Evol 45(3):203–217. https://doi.org/10.1016/s0047-2484(03)00106-4

Soulé ME, Estes JA, Berger J, Del Rio CM (2003) Ecological effectiveness: conservation goals for interactive species. Conserv Biol 17(5):1238–1250. https://doi.org/10.1046/j.1523-1739.2003.01599.x

Thalmann OB, Shapiro B, Cui P, Schuenemann VJ, Sawyer SK, Greenfield DL, Germonprè MB, Sablin MV, Lopez-Giraldez F, Domingo-Roura X, Napierala H, Uerpmann H-P, Loponte DM, Acosta AA, Giemsch L, Schmitz RW, Worthington B, Buikstra JE, Druzhkova A, Graphodatsky AS, Ovodov ND, Wahlberg N, Freedman AH, Schweizer RM, Koepfli K-P, Leonad JA, Meyer M, Krause J, Paabo S, Green RE, Wayne RK (2013) Complete mitochondrial genomes of ancient canids suggest a European origin of domestic dogs. Science 342(6160): 871–874. https://doi.org/10.1126/science.1243650

van der Valk T, Pečnerová P, Díez-del-Molino D, Bergström A, Oppenheimer J, Hartmann S, Xenikoudakis G, Thomas JA, Dehasque M, Sağlıcan E, Fidan FR, Barnes I, Liu S, Somel M, Heintzman PD, Nikolskiy P, Shapiro B, Skoglund P, Hofreiter M, Lister AM, Götherström A, Dalén L (2021) Million-year-old DNA sheds light on the genomic history of mammoths. Nature 591:265–269. https://doi.org/10.1038/s41586-021-03224-9

Weiss A, Kroeger T, Haney JC, Fascione N (2007) Social and ecological benefits of restored wolf populations. The 72nd north American wildlife and natural resources conference, Portland, OR. Volume: transactions of the 72nd north American wildlife and natural resources conference. www.istat.it/it/archivio/200929

The manufacturer's authorised representative in the EU is Springer Nature Customer Service Centre GmbH, Europaplatz 3, 69115 Heidelberg, Germany. If you have any concerns regarding our products, please contact ProductSafety@springernature.com

Printed and bound by CPI Group (UK) Ltd, Croydon, CR0 4YY
22/09/2026
02221055-0017